AF576024

Smart Energy and Intelligent Transportation Systems

Smart Energy and Intelligent Transportation Systems

Editors

Albert Y. S. Lam
Bogusław Łazarz
Grzegorz Peruń

MDPI • Basel • Beijing • Wuhan • Barcelona • Belgrade • Manchester • Tokyo • Cluj • Tianjin

Editors
Albert Y. S. Lam
The University of Hong Kong
Hong Kong

Bogusław Łazarz
Silesian University of Technology
Poland

Grzegorz Peruń
Silesian University of Technology
Poland

Editorial Office
MDPI
St. Alban-Anlage 66
4052 Basel, Switzerland

This is a reprint of articles from the Special Issue published online in the open access journal *Energies* (ISSN 1996-1073) (available at: https://www.mdpi.com/journal/energies/special_issues/energy_intelligent_transportation).

For citation purposes, cite each article independently as indicated on the article page online and as indicated below:

LastName, A.A.; LastName, B.B.; LastName, C.C. Article Title. *Journal Name* **Year**, *Volume Number*, Page Range.

ISBN 978-3-0365-4453-3 (Hbk)
ISBN 978-3-0365-4454-0 (PDF)

Contents

Editorial

Smart Energy and Intelligent Transportation Systems

Albert Y. S. Lam [1,2,*], Bogusław Łazarz [3] and Grzegorz Peruń [3]

1 Fano Labs, Hong Kong, China
2 Department of Electrical and Electronic Engineering, The University of Hong Kong, Hong Kong, China
3 Department of Road Transport, Faculty of Transport and Aviation Engineering, Silesian University of Technology, 40-019 Katowice, Poland; boguslaw.lazarz@polsl.pl (B.Ł.); grzegorz.perun@polsl.pl (G.P.)
* Correspondence: albert@fano.ai or ayslam@eee.hku.hk

1. Introduction

With the Internet of things and various information and communication technologies, a city can manage its assets in a smarter way, constituting the urban development vision of a smart city. This facilitates a more efficient use of physical infrastructure and encourages citizen participation. Smart energy and smart mobility are among the key aspects of the smart city, in which the electric vehicle (EV) is believed to take a key role. EV adoption in the market can clearly provide supporting evidence. When comparing to 2019, the 2020 global EV stock showed a 43 % increase, hitting the 10 million mark.

EVs are powered by various energy sources or the electricity grid. With proper scheduling, a large fleet of EVs can be charged from charging stations and parking infrastructures. Although the battery capacity of a single EV is small, an aggregation of EVs can perform as a significant power source or load, constituting a vehicle-to-grid (V2G) system. V2G refers to a system allowing EVs to communicate with the power system for demand response services by either discharging their excessive energy to the grid or by being charged with the excessive electricity from the grid. Besides acquiring energy from the grid, in V2G, EVs can also support the grid by providing various demand response and auxiliary services. We can reduce our reliance on fossil fuels and utilize renewable energy more effectively. V2G enables EVs to store electricity generated from renewable energy sources so as to overcome the intermittency of renewable due to different weather and time of day conditions.

The EV market is growing very quickly, and there will likely be an abundance of EVs running on the road in the near future. EVs are also important building blocks to developing intelligent transportation systems. The self-control of autonomous vehicles (AVs) and the systematic remote control of AV fleets will bring smart energy and intelligent transportation systems into new dimensions. We can develop a public transportation system with AVs, in which a fleet of AVs is managed to accommodate transportation requests, offering point-to-point services with ride sharing. AVs can also participate in V2G to support various V2G services. Through properly coordinating AVs in appropriate parking facilities, it has been shown that AVs can facilitate better V2G services with higher efficiency. On the other hand, an energy delivery system can be built upon the transportation network and EVs utilized as energy carriers to transport energy over a large geographical region. With proper routing, energy can be transmitted from sources to destinations as in a packet-switched network. Such a system can complement the power network and enhance the overall power system performance.

This Special Issue contributes to the smart energy and intelligent transportation system agenda through enhanced scientific and multi-disciplinary knowledge intended to improve performance and deployment by bringing some focus to electric and autonomous vehicles in order to meet technical, socio-economic, and environmental goals, as well as for energy security. We are particularly interested in investigating how smart energy technologies contribute to intelligent transportation systems, and vice versa.

Citation: Lam, A.Y.S.; Łazarz, B.; Peruń, G. Smart Energy and Intelligent Transportation Systems. *Energies* **2022**, *15*, 2900. https://doi.org/10.3390/en15082900

Received: 8 March 2022
Accepted: 12 April 2022
Published: 15 April 2022

2. Summary of the Contributions

Research works by scholars from divergent backgrounds are included in this Special Issue. Refs. [1,2] focus on the health of the vehicle and road–rail accidents. Ref. [3] investigates the control of charging in V2G while [4] develops an EV recommendation system. Ref. [5] studies the loading hub placement in an electric bicycle-oriented city logistic system.

Ref. [1] makes use of artificial neural networks in diagnosing the technical condition of driving systems operating under variable conditions, in which the effects of temperature and load variations on the values of diagnostic parameters were taken into consideration. A new approach is proposed to train the network using a learning set from the efficient system only. The responses to new data from the undamaged system are compared to the response to data recorded for three damage states: misalignment, unbalance, and simultaneous misalignment. As a normalized measure of the deviations, a diagnostic parameter for the faulted system is derived from the result for the undamaged condition.

Ref. [2] focuses on assessing the likelihood of the occurrence of various effects of road–rail accidents in different selected situations.The specificity of the road–rail accidents requires a separate characteristic to categorize types of incidents and to specify the affecting factors with an assessment of the strength of this impact. Classification trees are adopted to deal with the ordinal form of the dependent variables. The experiment results facilitate the characterization and assessment of the danger and constitute guidelines for taking preventive actions.

The massive adoption of EVs creates an unprecedented energy load for the power system, in which the stability and quality are compromised by multiple simultaneously connected vehicles, especially on a local distribution level. In [3], a choice-based pricing algorithm is proposed to indirectly control the charging and V2G activities of EVs in non-residential facilities. Two metaheuristic methods are applied to solve the proposed optimization problem with a comparative analysis for performance evaluation.

Due to the Act on Electromobility and Alternative Fuels, EVs are becoming popular in Poland, in which local government units and state administration are expanding their EV fleets. The expansion of the fleet should be well-planned and supported by economic analyses. Ref. [4] develops an EV recommendation system which meets the needs of the local and state administration to the greatest extent. A multi-criteria decision analysis method is designed with the Monte Carlo method, and it allows of promoting more sustainable vehicles with high technical, economic, environmental and social parameters.

Electric cargo bicycles are popular transport for last-mile goods deliveries in urban areas with restricted traffic. In a cargo bike delivery system, loading hubs serve as intermediate points between vans and bikes ensuring loading, storage, and e-vehicle charging operations. The loading hub placement a key problem for designing city logistic systems, which heavily rely on electric bicycles. In [5], the authors propose a mathematical model by considering consignees and loading hubs as vertices in the graph constituting a transport network. This allows determination of the location of a loading hub under stochastic demands for transport services in the closed urban area.

Author Contributions: All authors contributed equally to the conceptualization and realization of the manuscript. All authors have read and agreed to the published version of the manuscript.

Funding: This research received no external funding.

Conflicts of Interest: The authors declare no conflict of interest.

References

1. Pawlik, P.; Kania, K.; Przysucha, B. The Use of Deep Learning Methods in Diagnosing Rotating Machines Operating in Variable Conditions. *Energies* **2021**, *14*, 4231. [CrossRef]
2. Kozłowski, E.; Borucka, A.; Świderski, A.; Skoczyński, P. Classification Trees in the Assessment of the Road–Railway Accidents Mortality. *Energies* **2021**, *14*, 3462. [CrossRef]

3. Latinopoulos, C.; Sivakumar, A.; Polak, J.W. Optimal Pricing of Vehicle-to-Grid Services Using Disaggregate Demand Models. *Energies* **2021**, *14*, 1090. [CrossRef]
4. Ziemba, P. Multi-Criteria Stochastic Selection of Electric Vehicles for the Sustainable Development of Local Government and State Administration Units in Poland. *Energies* **2020**, *13*, 6299. [CrossRef]
5. Naumov, V. Substantiation of Loading Hub Location for Electric Cargo Bikes Servicing City Areas with Restricted Traffic. *Energies* **2021**, *14*, 839. [CrossRef]

Article

The Use of Deep Learning Methods in Diagnosing Rotating Machines Operating in Variable Conditions

Paweł Pawlik [1,*], Konrad Kania [2] and Bartosz Przysucha [2]

[1] Department of Mechanics and Vibroacoustics, Faculty of Mechanical Engineering and Robotics, AGH University of Science and Technology, al. A. Mickiewicza 30, 30-059 Kraków, Poland
[2] Department of Quantitative Methods in Management, Faculty of Management, Lublin University of Technology, ul. Nadbystrzycka 38 D, 20-618 Lublin, Poland; k.kania@pollub.pl (K.K.); b.przysucha@pollub.pl (B.P.)
* Correspondence: pawlik@agh.edu.pl

Abstract: This paper presents the use of artificial neural networks in diagnosing the technical condition of drive systems operating under variable conditions. The effects of temperature and load variations on the values of diagnostic parameters were considered. An experiment was conducted on a testing rig where a variable load was introduced corresponding to the load of the main gearbox of the bucket wheel excavator. The signals of vibration acceleration on the gearbox body, rotational speed, and current consumption of the drive motor for different values of oil temperature were measured. Synchronous analysis was performed, and the values of order amplitudes and the corresponding values of current, speed, and temperature were determined. Such datasets were the learning vectors for a set of artificial deep learning neural networks. A new approach proposed in this paper is to train the network using a learning set consisting only of data from the efficient system. The responses of the trained neural networks to new data from the undamaged system were performed against the response to data recorded for three damage states: misalignment, unbalance, and simultaneous misalignment and unbalance. As a result, a diagnostic parameter as a normalized measure of the deviation of the network results was developed for the faulted system from the result for the undamaged condition.

Keywords: condition monitoring; vibroacoustic diagnostics; gearbox; power transmission systems; neural networks; deep learning

Citation: Pawlik, P.; Kania, K.; Przysucha, B. The Use of Deep Learning Methods in Diagnosing Rotating Machines Operating in Variable Conditions. *Energies* **2021**, *14*, 4231. https://doi.org/10.3390/en14144231

Academic Editors: Davide Astolfi and Bogusław Łazarz

Received: 20 June 2021
Accepted: 9 July 2021
Published: 13 July 2021

1. Introduction

Monitoring systems using vibration signals are often used in the industry to diagnose the technical condition of machinery. Systems based on parameters such as RMS (root mean square), crest factor, and peak of vibration signal are perfect for machines operating in constant operating conditions, where one can assume threshold values for these parameters, whereby exceeding them signals a system fault. However, these parameters are not sufficient to diagnose machines operating under variable loads caused by varying operating conditions [1]. Variable loading causes a change in speed, which makes diagnosis difficult using classical methods based on spectral analysis of the time signal. Synchronous methods are used in the diagnosis of machines operating at variable speed, which are based on the synchronization of the signal carrying diagnostic information with the rotational speed of the tested object [2–5]. However, varying operating conditions such as load or operating temperature can also affect the amplitude of the diagnostic signal [6–10]. An increase in amplitude due to a variable load can be misinterpreted by monitoring systems. Therefore, this impact must be considered in the monitoring process. A previous study [1] investigated the effect of load and speed on the amplitude values of the characteristic components and proposed a method for scaling these parameters. It is also possible to find papers on diagnosing a planetary gearbox operating under variable conditions, which addressed the problem of changing parameter values from load [7,11]. The solution to this problem can be realized using artificial neural networks [12–15]. The authors of [16] presented a one-dimensional multi-scale domain adaptive network in bearing diagnostics for different degrees of load. The preliminary preparation of diagnostic signals is important in using neural networks to diagnose machines working in variable conditions. In [17], higher-order spectral analysis and multitask learning were used. However, in [18], a method of learning a convolutional network was proposed using a labeled date and pseudo-labeled date. The measurement of drive motor current in diagnosing gear damage and bearings [19–21] is increasingly being used. The presented cases used artificial neural networks as classifiers. In the learning process, vectors for the undamaged and damage states were used as inputs of the network. However, in industrial conditions, especially in the heavy industry, there are no measurement data registered during damage because not all machines are monitored, and the drive system components are produced in small amounts. Moreover, it is impossible to predict all types of faults and introduce them into the network learning process. Accordingly, we propose a solution based on learning neural networks with data recorded only during fault-free operation, which analyzes the responses of the trained networks to signals recorded when faults are introduced.

This paper presents the use of artificial neural networks in diagnosing the technical condition of drive systems operating under variable conditions. The effects of temperature and load variations on the values of diagnostic parameters were considered. An experiment was conducted on a testing rig, where a variable load was introduced, corresponding to the load of the main gearbox of the bucket wheel excavator. The vibration acceleration, speed, and current signals feeding the drive motor were recorded for different values of oil temperature. Synchronous analysis was performed, and the values of order amplitudes and the corresponding values of current, speed, and temperature were determined. Such datasets were the learning vectors for a set of artificial neural networks.

This article presents a new approach to diagnosing machines working in variable conditions. Instead of analyzing the order spectrum, an analysis of the amplitudes of orders as a function of changes in speed, current, and temperature was carried out. A new approach proposed in this paper is to train the network using a learning set consisting only of data from the efficient system. In the next step, the behavior of such a trained neural network on new data from the undamaged system was investigated in comparison with the response to data recorded during three fault conditions of the drive system. As a result, a diagnostic parameter as a normalized measure of the deviation of the network results was developed for the faulted system from the results for the fault-free condition.

2. Materials and Methods

2.1. Testing Facility Description

The testing rig (Figure 1) consisted of a TRAMEC EP 90/1 planetary gearbox driven by an electric motor controlled by a frequency converter. The load consisted of a second induction motor connected to the gearbox by a jaw coupling. The load motor was also controlled by a frequency converter, which allowed setting any load function of the system.

Vibration acceleration on the planetary gear case was measured using a PCB 356B08 triaxial acceleration sensor, the temperature was measured using an LM35 sensor, speed was measured using a laser tachometer, and electrical current was measured using an ACS714 sensor. All measurements of the experiment were obtained using a specially built measuring system based on the PXI platform (PCI Extension for Instrumentation). The platform comprised the PXI Trigger Bus, which allows measurement synchronization between individual measuring cards.

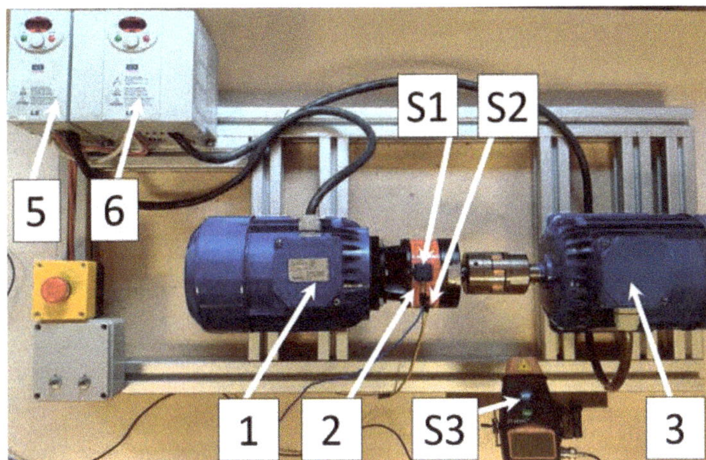

Figure 1. Laboratory bench, (**1**) drive motor, (**2**) planetary gearbox, (**3**) load motor, (**S1**) acceleration sensor, (**S2**) temperature sensor, (**S3**) tachometer, (**5**) frequency inverter for the drive motor, and (**6**) frequency inverter for the load motor.

2.2. Diagnostic Experiment

Measurements were made for four drive system conditions, which are shown in Table 1.

Table 1. Designation and measurement time for the different conditions of the drive system.

Marking	Machine Condition	Measurement Time
$F0$	Undamaged	30 min
$F1$	Misaligned	30 min
$F2$	Unbalanced	30 min
$F3$	Misaligned and unbalanced	30 min

The misalignment condition ($F1$) consisted of placing 0.5 mm thick shims under the front feet of the drive motor. The unbalance condition ($F2$) was introduced by placing additional mass (13 g) on the output shaft coupling. Condition $F3$ consisted of the simultaneous introduction of misalignment and unbalance. Signals of 30 min duration were recorded for each condition. Measurements were carried out when warming up a system for a temperature in the range of 35–40 °C.

The variable load changes from 1.3 Nm to 4.0 Nm were entered into the system. This caused a change in the rotational output shaft speed of 730–758 RPM and in the RMS value of drive motor current in the range of 2.48–2.51 A. The system was subjected to a variable load corresponding to the load occurring on the main gear of the bucket wheel excavator. The reference signal was obtained from the KWK 1500 s excavator main gear monitoring system, the subject of the patent in [22]. The main gear of the bucket wheel excavator is a bevel-planetary gearbox that drives a bucket wheel. The input of the bucket in the ground causes the system load. With a fixed set voltage value on the drive engine, the load causes a change in speed and vibration amplitude. Diagnosing the main gear of the wheel excavator has already been addressed in [7,11,23], which investigated the signals coming from industrial conditions. However, in industrial conditions, a controlled experiment with given damage cannot be carried out. Therefore, the experiment was carried out in laboratory conditions. The variable load-induced velocity change signals recorded under industrial conditions were scaled up to the capabilities of the laboratory bench.

2.3. The Method of Determining Diagnostic Parameters

The proposed signal analysis method is based on the order analysis algorithm. The order spectrum is determined using a method based on resampling the vibration acceleration signal against the input shaft speed. Figure 2 shows the schematic of the order analysis algorithm. In the first phase, the tachometer signal is subjected to an interpolation procedure using a cascaded integrator–comb filter (CIC). Next, on the basis of filtered tachometer signal, a vibration signal resampling procedure is performed to determine the vibration signal against the rotation angle (even angle signal). In the resampling method, time samples are converted to angle samples. The time samples are samples of the physical signal that are equally spaced in time. The angle samples are samples that are equally spaced in the rotation angle. The signal resampled as such can be subjected to a fast Fourier transform (FFT), the result of which is an order spectrum. An order spectrum represents amplitude as a function of order rather than as a function of frequency. The orders correspond to multiples of the rotational frequency of the shaft on which the speed measurement is performed [24]. In the case under consideration, the rotational speed measurement is carried out at the output shaft of the gearbox.

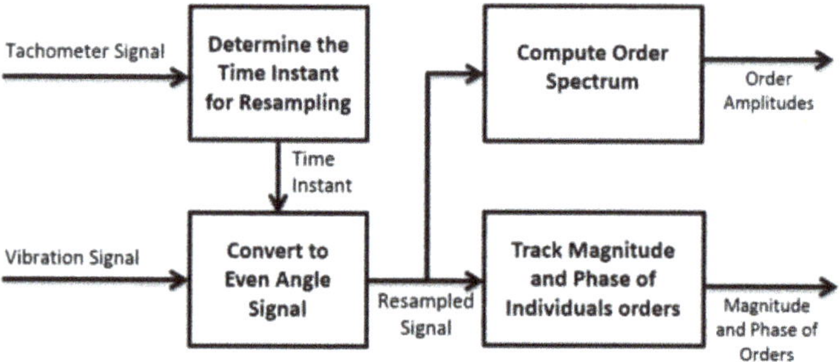

Figure 2. Schematic of the order analysis algorithm [24].

In addition to the averaged order spectrum, the time course of the amplitudes of the individual orders can be obtained from the order analysis. Information about the technical condition of the tested object can be obtained by monitoring the amplitudes of the characteristic orders. However, the change in amplitudes can also be caused by a change in the load on the system [6,25]. The load can be measured indirectly by measuring the electrical current consumption of the drive system. However, it is not always possible to accurately measure electrical current in industrial conditions, especially a measurement synchronized with a vibration measurement. However, in industrial settings, the driving motor is often supplied with a constant voltage and frequency, and any variation in speed is due to a change in load. The relationship between load change and speed recorded on the testing rig is shown in Figure 3. The rotational speed decreased as a result of the increase in the system load, while the load increase resulted in a higher power consumption by the driving motor.

In the test object, the rotational speed was determined by measuring the keyphasor signal. This measurement was synchronized with the vibration acceleration measurement, allowing load variations to be included in diagnostic estimates, such as characteristic component amplitudes. Figure 4 shows the speed waveform of the input shaft of the diagnosed object during operation under load. Speed changes were caused by changes in the system load because the frequency and voltage applied to the motor driving the system were constant. A similar case can be encountered in industrial settings. In the experiment conducted, rotational speed variation waveforms recorded on the main gear drive system of a bucket wheel excavator were used. The shape of the rotational speed variation waveform shown in Figure 4 corresponds to the waveform recorded on the main gear of a wheeled excavator operating in a brown coal mine.

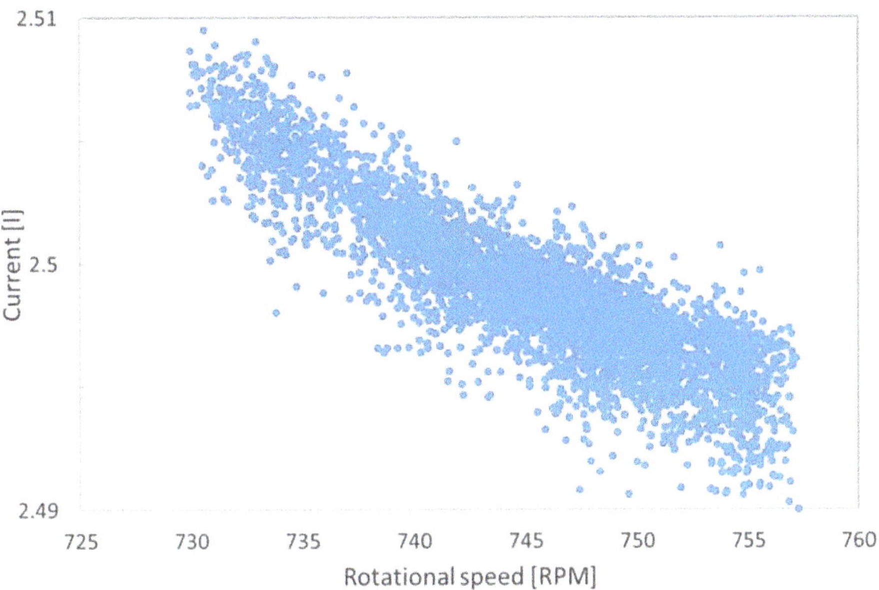

Figure 3. The relationship between rotational speed and current consumption of the drive motor.

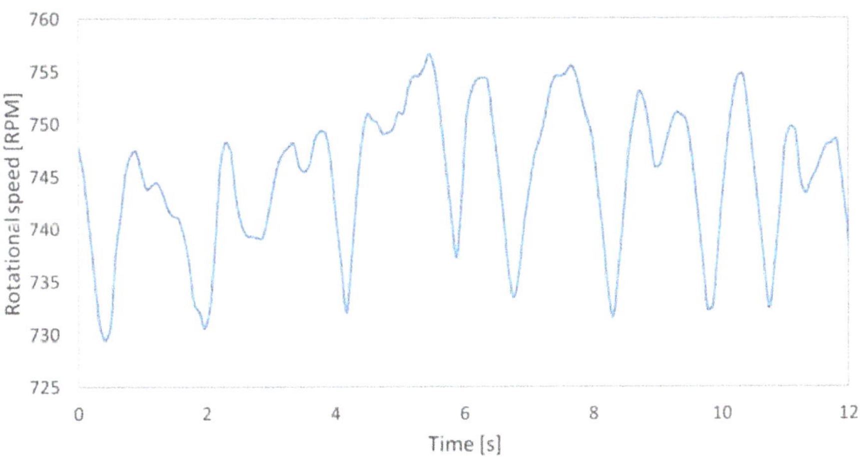

Figure 4. Input shaft rotational speed waveform.

Changes in rotational speed associated with varying load resulted in a significant change in the amplitude of the characteristic orders. The value of the order amplitude for the loaded and unloaded system changed significantly, hindering the diagnosis based on this parameter. Figure 5 shows the meshing order amplitude (no. 72) as a function of rotational speed for the efficient system, which varied under load. For maximum rotational speed value, a small load was applied to the system. As the load increased, the speed decreased, while the value of the meshing order amplitude increased because of the interaction between the teeth increasing.

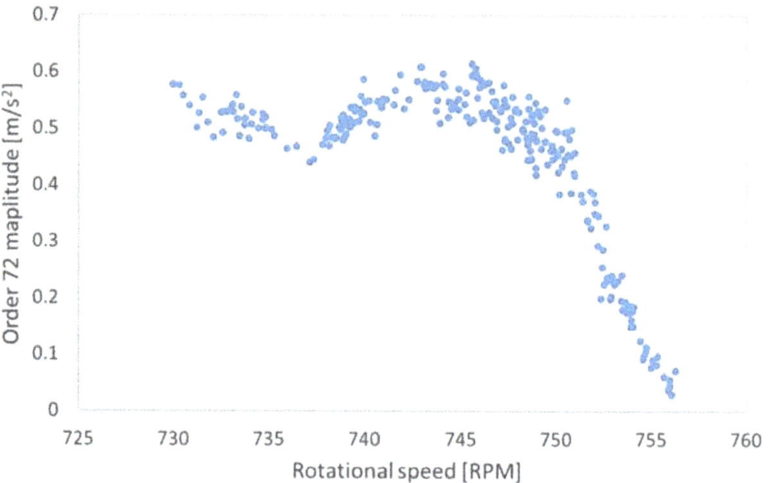

Figure 5. Dependence of order no. 72 amplitude as a function of speed.

The values of ordinate amplitudes are also affected by temperature, and this was shown in [9]. Diagnostics should be performed for a constant temperature or by considering its influence during inference. Therefore, temperature waveforms measured on the gearbox body were also recorded.

Changes in amplitudes of characteristic orders in the order spectrum are usually analyzed in vibroacoustic diagnostics. According to the literature [6,26–28], individual faults or installation defects are characterized by changes in the amplitudes of the corresponding orders, e.g., a change in the amplitude of order no. 1 is symptomatic of unbalance, while a change in the order corresponding to the number of claws of the jaw coupling is due to system misalignment. Therefore, each order obtained from the order spectrum was considered separately because the characteristics of the order amplitudes as a function of load (rotational speed) would change with failure for each order in a different way [25].

As a result of the order analysis, 100 consecutive orders of vibration acceleration recorded in the direction parallel and perpendicular to the shaft axis were obtained. Current, temperature, and speed were recorded simultaneously. Signals of 30 min duration were recorded for each condition. Next, all the recorded signals were divided into time frames of 30 s each. Sets of signals of this length were sequentially subjected to the processing algorithm shown in Figure 6. The processing time of 30 s signals was shorter than 30 s; therefore, the algorithm can be implemented in continuous monitoring systems.

This procedure was performed for data recorded during correct operation (*F*0) and for signals recorded for individual faults *F*1, *F*2, and *F*3.

First, the vibration signal was subjected to order analysis. The results of this analysis were the waveforms of order amplitudes and speed over time. Figure 6 shows the case for one order. The waveforms of order amplitude, speed, temperature, and current were then sorted in ascending order with respect to speed. The results of such sorting were the courses of these parameters relative to the rotation speed. The next step was to determine the moving average for *N* consecutive elements. Averaging was used to reduce data scatter. The data thus prepared were input vectors for artificial neural networks.

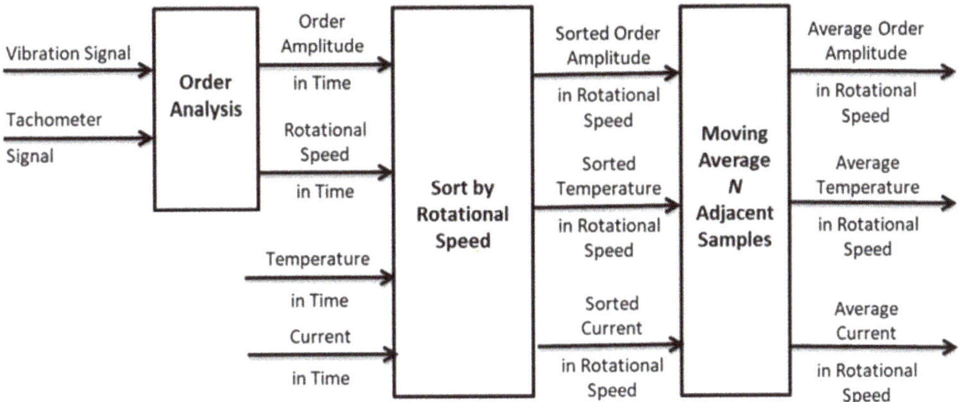

Figure 6. Measurement signal processing algorithm for a single order.

2.4. Using the Artificial Neural Network for Condition Assessment

Deep learning artificial neural networks were used to evaluate the condition of the diagnosed drive train. The idea behind the method was to design the learning process of the neural network such that it detects abnormal situations in systems where faults have not historically occurred or have occurred so infrequently that typical classifiers based on supervised learning could not be used. Accordingly, only data from the undamaged system were used for the learning process.

In addition, it was considered that neural network learning is more efficient when there are significantly fewer outputs than the input data variables [29–31]. For this reason, instead of one model evaluating all 100 orders simultaneously, a separate model was built for each order.

From a diagnostic perspective, the best solution would be to reduce the diagnostic decision to a single numerical parameter. Therefore, the system was designed to produce a synthetic diagnostic parameter for each order by exploiting the ability of deep neural networks to recognize patterns autonomously [32,33]. Moreover, some theoretical results suggested that network-based systems can lead to proper robustness [34].

The input data vector for each network were $x_k \in R^5$, with its components including the following values, in respective order:

- output shaft rotational speed;
- oil temperature;
- the current drawn by the motor;
- amplitude values of order k in the x-axis;
- z-axis amplitude values of the order k;

The input layer, therefore, consisted of five neurons, followed by two hidden layers [35,36]:

- a 30-neuron layer with RELU activation function and a drop-out layer with a drop rate of 20%;
- a 20-neuron layer with RELU activation function and a drop-out layer with a drop rate of 10%.

The output layer of the individual networks consisted of a single neuron returning a numerical value, which, for the correct state of the machine ($F0$), should return values close to the constant value determined in the network learning stage. In the experiment conducted, the conventional value of this constant was taken as $m = 1$.

The expected effect of modeling was that the applied neural network should realize some continuous transformation [32] $H_k : R^5 \to R$ with the property described in Equation (1).

$$H_{k,i}(x_{k,i}) = \begin{cases} m + \varepsilon \ \ when \ x_{k,i} \in F0 \\ t(x_{k,i}) \ \ when \ x_{k,i} \notin F0 \end{cases}, \quad (1)$$

where $t(x_{k,i})$ is the function from R^5 in R with the property that $|m + \varepsilon - t(x_k)|$ is significantly different from 0, $x_{k,i} \in F0$ is the vector containing data from the undamaged system for order k, $x_{k,i} \notin F0$ is the vector containing data from the faulty system ($F1$, $F2$, or $F3$) for order k, i is the index of the input vector, $m = 1.00$, and $\varepsilon \sim N(0, \sigma_0^2)$.

This means that, if $x_{k,i} \in F0$, then the value of the function $H_{k,i}$ should be close to the value of $m = 1$. However, if the vector $x_{k,i}$ corresponds to the operation of the defective transmission, the neural network should return a value different enough from m such that, taking into account the amount of variance, σ_0^2 this difference could be detected by statistical analysis. Since the transformation in Equation (1) is continuous, this guarantees the stability of the network (similar values of the network response come from similar input values). There are some papers where similar observations were made for obtaining stabilization of randomized systems [37].

The primary task in constructing such an architecture is to avoid network overfitting. Due to the fact that we only used the data of a well-functioning machine during the network learning process, such a network may tend to adjust the weights in such a way that, regardless of the output, it returns a response equal to the learned value (i.e., returning a parameter close to m regardless of the input state). Therefore, the network training process used as input a set of vectors $x_{k,i}$ associated with the state $F0$. On the other hand, random values from the distribution $N(m = 1, \sigma = 0.1)$ were generated as the output vector. Adding a perturbation with a small variance to the output vector, as early as at the network learning stage, reduced the risk of trivializing the function realized by the neural network.

Further reducing the risks associated with overfitting involved the following steps:

- using drop-out layers in the network architecture and randomizing the network by randomly deactivating neurons during learning;
- the learning set was randomly divided into a learning set and a validation set (which the network did not formally use for learning).

In addition, the use of these operations renders the network training process nondeterministic, such that the process of training the network for each order can be repeated independently, each time obtaining slightly different values of the weights in each layer, for the same input data. One can notice that this feature of the training process can be exploited to implement the bagging technique. The learning process was, therefore, repeated 120 times, resulting in 120 network models for each order. This approach provided a key mechanism to reduce the risk of system overfitting, as it dispersed this risk across multiple, independent models. In order to achieve that dispersion, evaluation values for 120 models were considered instead of considering the evaluation of a single model. The algorithm of the learning process is shown in Figure 7.

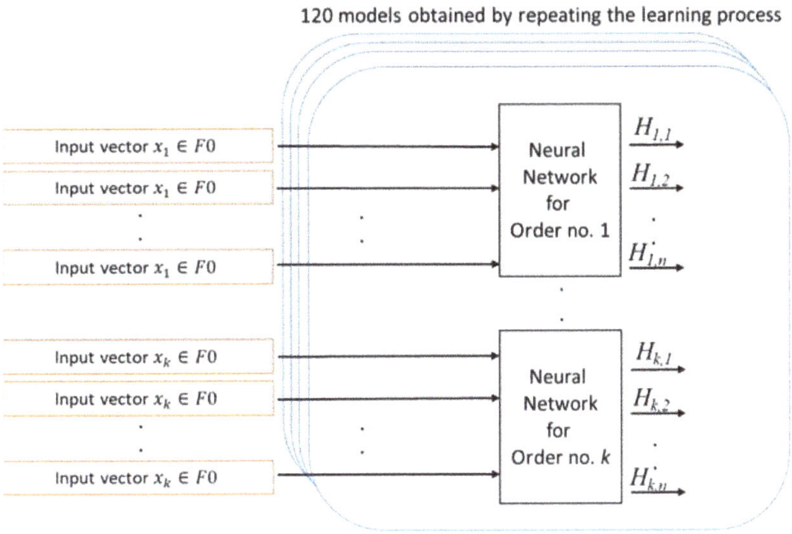

Figure 7. Learning process algorithm.

2.5. Verification of Artificial Neural Network Functionality

The functionality of the obtained model was verified after the network learning process. The responses of the proposed solution were investigated when the vectors obtained from the measurements for the machine fault conditions $F1$, $F2$, and $F3$ and for the correct operation state $F0$ were input. Vectors that were not used during network learning were used to verify the undamaged conditions.

The network response values $H_{k,i}$ were determined for all orders, $k \in \{1...100\}$. Next, the average value was determined $\hat{H}_k$ for each order according to the following relationship:

$$\hat{H}_k = \frac{1}{n}\sum_{i=0}^{n} H_{k,i}(x_{k,i}), \tag{2}$$

where n is the number of input vectors, i is the index of the input vector, and $H_{k,i}(x_{k,i})$ is the i-th output value of the network for the k-th order.

This process was then repeated for 120 trained models. The 120 values of the parameter $\hat{H}_k$ were obtained for each order. This process was repeated for each condition of the tested drive train. The process of analyzing the functionality of the learned neural network for the $F1$ condition is shown in Figure 8.

For verification, ratings for states with damage were compared to ratings for a defect-free system. Verification of whether the average of 120 evaluations $\hat{H}_k$ for each order differed significantly from the average of 120 evaluations $\hat{H}_0$ for condition $F0$ was conducted using an ANOVA test.

Two independent input datasets of the undamaged system were used when testing the response for the undamaged system. This approach allowed not only determining whether there was a statistically significant difference between the network response to the fault condition and the response to the no-fault condition, but also measuring the magnitude of the observed discrepancy in a normalized way using the parameter ω^2. According to the recommendations [38], negative values of this coefficient, which occur when the size of the noise exceeds the size of the effect, were treated as zeros. Taking this correction into account, the value ω^2 is described by Equation (3).

$$\omega^2 = \begin{cases} \frac{SS_{effect} - df_{effect} \cdot MS_{error}}{MS_{error} + SS_{total}}, & when\ SS_{effect} > df_{effect} \cdot MS_{error} \\ 0, & otherwise \end{cases}, \tag{3}$$

where SS_{effect} is the sum of squares of the effect from the ANOVA test, df_{effect} is the number of degrees of freedom of the effect from the ANOVA test, SS_{total} is the total sum of squares from the ANOVA test, and MS_{error} is the mean squared error of the residuals from the ANOVA test.

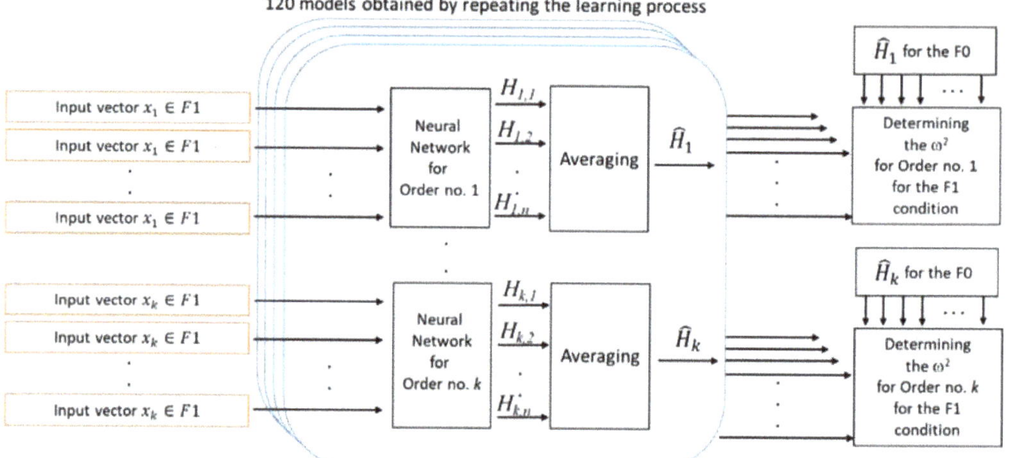

Figure 8. Algorithm of neural network functional analysis process for the *F*1 condition.

The parameter ω^2 allows a numerical evaluation normalized to the interval [0, 1] of the deviation of the results obtained from the neural networks for a given technical condition from the result for the condition without faults. The values of the parameter ω^2 can be classified by determining the power of the effect on the Cohen scale [39] (Table 2).

Table 2. Cohen Scale.

Range of ω^2	The Power of the Effect
0–0.1	No effect
0.1–0.3	Little effect
0.3–0.5	Moderate effect
0.5–1.0	Large effect

3. Results and Discussion

3.1. Results of the Order Analysis

When analyzing the spectrum of orders of vibration acceleration signals recorded during machine operation at variable load (Figure 9and Figure 11), a significant standard deviation can be observed for the gear coupling band (orders around coupling order no. 72). This scatter is caused by the high dynamics of the load (Figure 4) applied to the planetary gearbox.

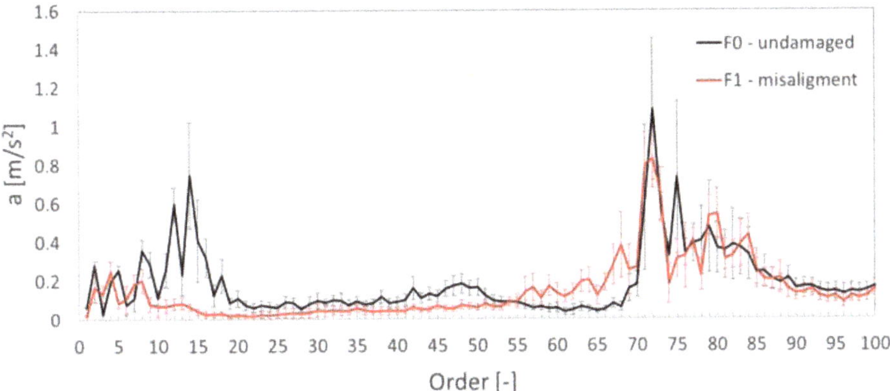

Figure 9. Order spectra of vibration acceleration signal (vertical direction) for the undamaged system (**black**) and for the misaligned system (**red**).

Figure 9 shows the spectra of vibration acceleration orders (vertical direction) for the system without damage (black) and with the introduced misalignment (red).

In the case of misalignment, one would expect an increase in the amplitude of order no. 4 corresponding to the number of clutch teeth. However, there is a noticeable increase in amplitude near the coupling band—orders no. 60 to 70. On the other hand, observing the dependence of the amplitude of the order no. 4 as a function of rotational speed (Figure 10), we can observe more than a twofold increase for the speed equal to 745 RPM and for the speed over 755 RPM. Moreover, in the range of 748–755 RPM, the values for the amplitudes for the misaligned system are smaller than those for the efficient system. On the order spectrum, the difference is smaller because the order spectrum represents a value averaged over a period of over a dozen rotations.

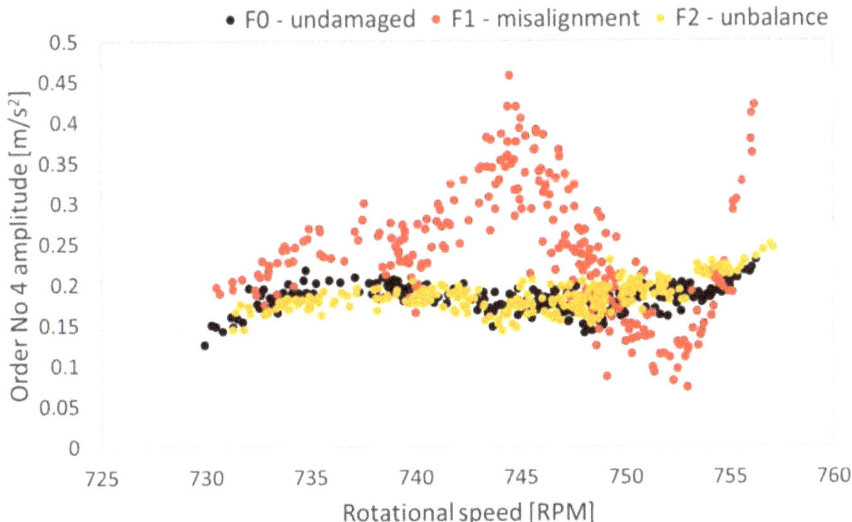

Figure 10. Order no. 4 amplitudes of vibration acceleration signal (vertical direction) versus rotational speed for the system without damage (**black**), for the misaligned system (**red**), and for the unbalanced system (**orange**).

Comparing the order spectrum from the undamaged system with the order spectrum from the imbalanced system, it is difficult to see the increase in order no. 1 amplitude, which is symptomatic of this damage (Figure 11). These differences are relatively small when compared to the values of the coupling order amplitudes. However, the order no. 1 has a much smaller standard deviation than the orders in the coupling band of Figure 11.

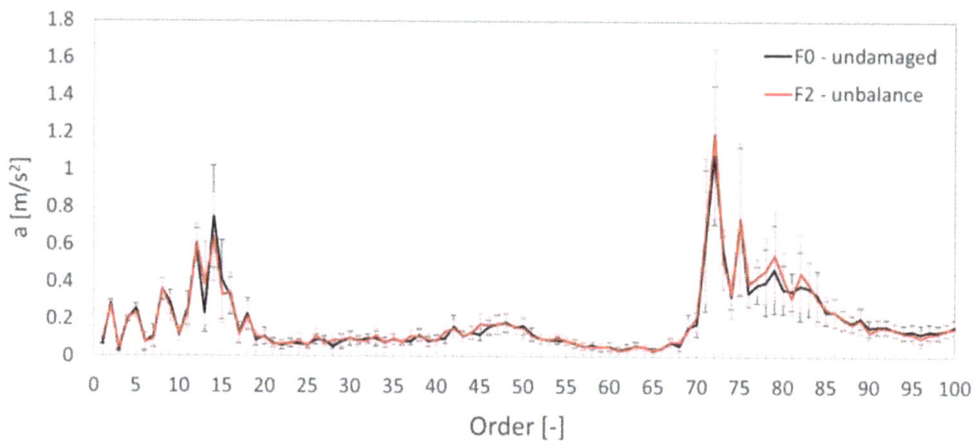

Figure 11. Order spectra of vibration acceleration signal (vertical direction) for the undamaged system (**black**) and for the unbalanced system (**red**).

A much smaller scatter of amplitudes of the order no. 1 for unbalance than for misalignment is also seen in Figure 12. However, the amplitude for the alignment signal increased for the entire speed range by a value of 0.04 m/s^2, constituting more than 70% of the initial value.

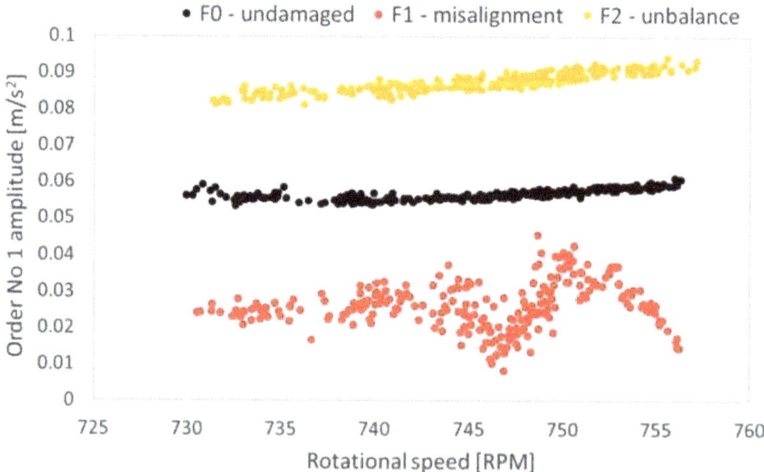

Figure 12. Order no. 1 amplitudes of vibration acceleration signal (vertical direction) versus rotational speed for the system without damage (**black**), for the misaligned system (**red**), and for the unbalanced system (**orange**).

In the case of variable load with a dynamic character causing speed changes shown in Figure 4, the analyzed damage is hardly visible in the order spectrum. Significant changes are visible only after careful analysis of the dependence of the amplitudes of the individual orders on the changes in rotational speed, which are caused by the change in load. Therefore, it is necessary to take into account the changes in ordinate amplitudes due to loading in the diagnosis process.

Temperature also has a significant effect on the values of order amplitudes in the interlocking band. Figure 13 shows the order spectra of vibration acceleration for the signal recorded for different temperatures. There is a clear increase in order amplitudes occurring with increasing temperature. This confirms the need to consider the temperature in vibroacoustic diagnostics.

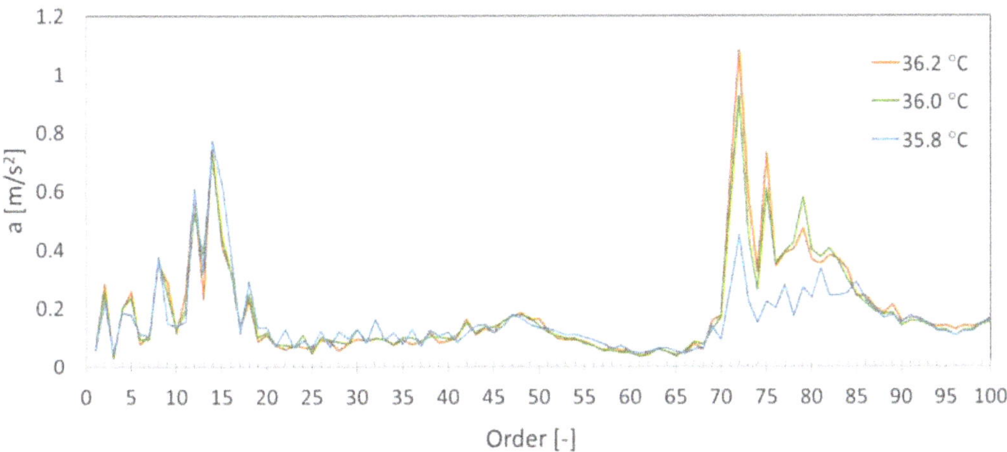

Figure 13. Order spectra of vibration acceleration signal (vertical direction) for a system without damage for different temperatures.

3.2. Results Obtained from Deep Learning Neural Networks

(Figures 14–16) show the values of the ω^2 parameter for each order. This parameter is a numerical evaluation of the deviation of the results obtained from the set of neural networks for a given fault from the results obtained for the no-fault condition.

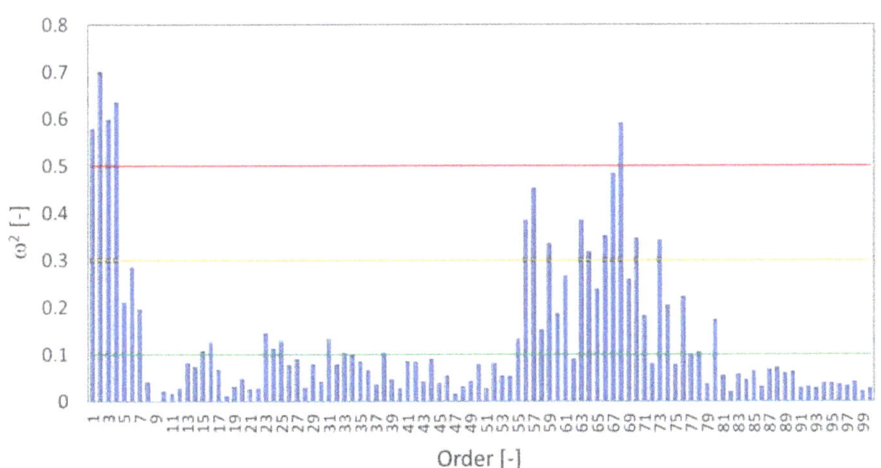

Figure 14. Values of the ω^2 parameter for individual orders—the system in the *F*1 condition (misalignment).

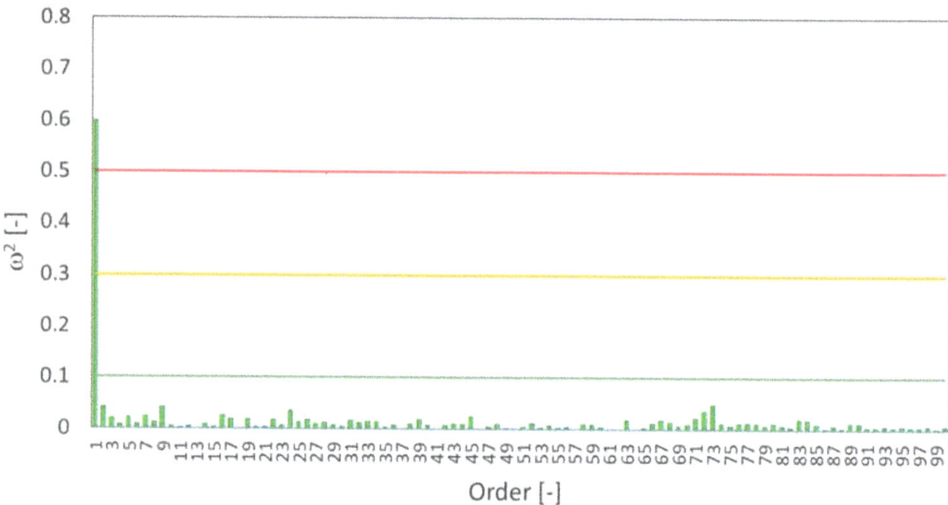

Figure 15. Values of the ω^2 parameter for particular orders—the system in the *F*2 condition (unbalanced).

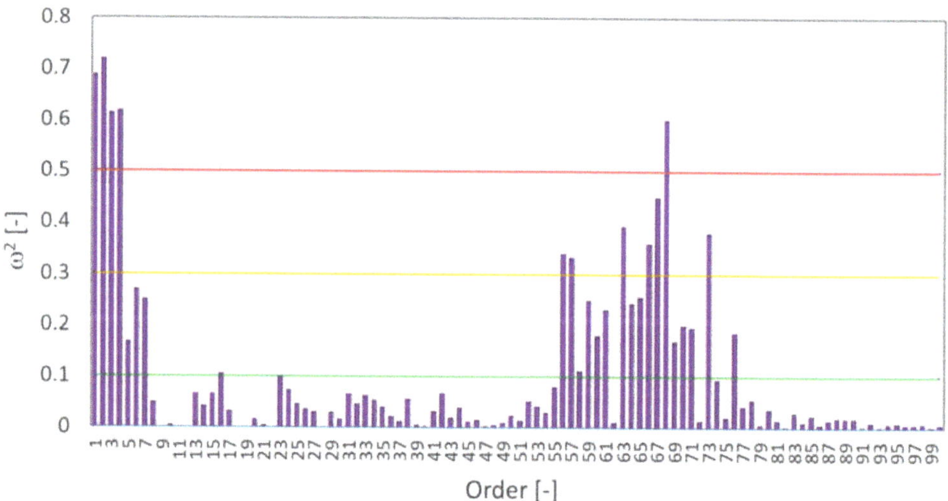

Figure 16. Values of the ω^2 parameter for individual orders—the system in the *F*3 condition (misalignment and unbalance).

First, the control data from the measurement for the undamaged system (*F*0) were provided to the input of the trained networks. These values for all orders were 0, indicating no effect according to the Cohen scale.

Figure 14 shows the values of the parameter ω^2 for the misalignment condition (*F*1). A large effect size can be observed for orders no. 1, 2, 3, 4, and 68, while a moderate effect can be observed for orders near the coupling order. The increase in the parameter for order no. 4 is most reasonable and is related to the number of teeth in the coupling used. On the other hand, large values of the ω^2 parameter for the coupling band may be due to increased inter-tooth interaction during system misalignment.

For the unbalanced condition (*F*2) (Figure 15), a large effect of the Cohen scale is observed only for order no. 1. The values of the ω^2 parameter for the other orders indicate no effect (differences between the response of the network to the condition *F*2 and *F*0). This is consistent with theory because unbalance affects the order no. 1 amplitude values with respect to the shaft on which it occurs.

In the case of the $F3$ condition (Figure 16), where both faults were introduced, similar values of the ω^2 parameter can be observed as in the $F1$ condition. However, the effect of unbalance is also evident by the increased amplitude for the parameter for order no. 1.

4. Conclusions

This paper presents the problem of diagnosing machines operating under variable conditions. A diagnostic experiment was conducted on a testing rig for diagnosing a planetary gearbox. The effects of varying load and oil temperature on vibration acceleration signals measured on the body of a planetary gear that was part of the drive system were analyzed.

For the applied variable load, the use of order spectrum was not sufficient to diagnose the introduced drive train damages. This paper shows that the analysis of the dependence of the order amplitudes on the changes in rotational speed (caused by varying load) allows observing significant changes.

A method to evaluate the drive train technical condition using a set of artificial deep learning neural networks was proposed. The input data to the network were values of order amplitudes, temperature, current, and corresponding values of rotational speed. A new approach was proposed to train a set of neural networks using only data recorded for a drive train without faults. The response of the network to given sets of data from a system with introduced faults, i.e., misalignment, unbalance, and simultaneous misalignment and unbalance, was then tested.

The ω^2 parameter was proposed to evaluate the deviation of the network results for introduced faults from the result for the no-fault condition. This parameter is determined for each order individually. Therefore, damage can be identified on the basis of previous diagnostic knowledge by observing the orders carrying information about each damage.

The use of artificial deep learning neural networks made it possible to take into account the effects of varying load and temperature on the values of the amplitudes of the characteristic orders.

Author Contributions: Conceptualization, P.P.; methodology, P.P., K.K. and B.P.; software, P.P. and K.K; validation, P.P., K.K. and B.P.; formal analysis, P.P. and K.K.; investigation, P.P.; resources, P.P. and K.K.; data curation, P.P. and K.K.; writing—original draft preparation, P.P., K.K. and B.P.; writing—review and editing, P.P., K.K. and B.P.; visualization, P.P.; supervision, P.P.; project administration, P.P.; funding acquisition, P.P. and B.P. All authors have read and agreed to the published version of the manuscript.

Funding: This research was funded by the Polish Ministry of Science and Higher Education, grant number 16.16.130.942, and by the Fund for Science and Research of Lublin University of Technology.

Institutional Review Board Statement: Not applicable.

Informed Consent Statement: Not applicable.

Data Availability Statement: Not applicable.

Conflicts of Interest: The authors declare no conflict of interest.

References

1. Pawlik, P.; Lepiarczyk, D.; Dudek, R.; Ottewill, J.R.; Rzeszuciński, P.; Wójcik, M.; Tkaczyk, A. Vibroacoustic Study of Powertrains Operated in Changing Conditions by Means of Order Tracking Analysis. *Eksploat. i Niezawodn. Maint. Reliab.* **2016**, *18*, 606–612. [CrossRef]
2. Randall, R.B. *Frequency Analysis*; Bruel & Kjær: Nærum, Denmark, 1987.
3. Wang, W.; Johnson, K.; Galati, T. Vibration Analysis of Planet Gear Bore-Rim Failure Using Enhanced Planet Time Synchronous Averaging. *Eng. Fail. Anal.* **2020**, *117*, 104942. [CrossRef]
4. Zhang, S.; Tang, J. Integrating Angle-Frequency Domain Synchronous Averaging Technique with Feature Extraction for Gear Fault Diagnosis. *Mech. Syst. Signal Process.* **2018**, *99*, 711–729. [CrossRef]

5. Cioch, W.; Knapik, O.; Leśkow, J. Finding a Frequency Signature for a Cyclostationary Signal with Applications to Wheel Bearing Diagnostics. *Mech. Syst. Signal Process.* **2013**, *38*, 55–64. [CrossRef]
6. Cempel, C. Application of TRIZ Approach to Machine Vibration Condition Monitoring Problems. *Mech. Syst. Signal Process.* **2013**, *41*, 328–334. [CrossRef]
7. Zimroz, R.; Bartkowiak, A. Two Simple Multivariate Procedures for Monitoring Planetary Gearboxes in Non-Stationary Operating Conditions. *Mech. Syst. Signal Process.* **2013**, *38*, 237–247. [CrossRef]
8. Peruń, G.; Łazarz, B. Modelling of Power Transmission Systems for Design Optimization and Diagnostics of Gear in Operational Conditions. *Solid State Phenom.* **2014**, *210*, 108–114. [CrossRef]
9. Pawlik, P. The Diagnostic Method of Rolling Bearing in Planetary Gearbox Operating at Variable Load. *Diagnostyka* **2019**, *20*, 69–77. [CrossRef]
10. Pawlik, P. The Use of the Acoustic Signal to Diagnose Machines Operated Under Variable Load. *Arch. Acoust.* **2020**, *45*, 263–270. [CrossRef]
11. Bartelmus, W.; Zimroz, R. A New Feature for Monitoring the Condition of Gearboxes in Non-Stationary Operating Conditions. *Mech. Syst. Signal Process.* **2009**, *23*, 1528–1534. [CrossRef]
12. Lipinski, P.; Brzychczy, E.; Zimroz, R. Decision Tree-Based Classification for Planetary Gearboxes' Condition Monitoring with the Use of Vibration Data in Multidimensional Symptom Space. *Sensors* **2020**, *20*, 5979. [CrossRef]
13. Popiołek, K.; Pawlik, P. Diagnosing the Technical Condition of Planetary Gearbox Using the Artificial Neural Network Based on Analysis of Non-Stationary Signals. *Diagnostyka* **2016**, *17*, 57–64.
14. Dabrowski, D. Condition Monitoring of Planetary Gearbox by Hardware Implementation of Artificial Neural Networks. *Meas. J. Int. Meas. Confed.* **2016**, *91*, 295–308. [CrossRef]
15. Łazarz, B.; Wojnar, G.; Czech, P. Early Fault Detection of Toothed Gear in Exploitation Conditions. *Eksploat. i Niezawodn. Maint. Reliab.* **2011**, *1*, 68–77.
16. Wang, K.; Zhao, W.; Xu, A.; Zeng, P.; Yang, S. One-Dimensional Multi-Scale Domain Adaptive Network for Bearing-Fault Diagnosis under Varying Working Conditions. *Sensors* **2020**, *20*, 6039. [CrossRef]
17. Hasan, M.J.; Sohaib, M.; Kim, J.M. A Multitask-Aided Transfer Learning-Based Diagnostic Framework for Bearings under Inconsistent Working Conditions. *Sensors* **2020**, *20*, 7205. [CrossRef]
18. Zhang, K.; Wang, J.; Shi, H.; Zhang, X.; Tang, Y. A Fault Diagnosis Method Based on Improved Convolutional Neural Network for Bearings under Variable Working Conditions. *Measurement* **2021**, *182*, 109749. [CrossRef]
19. Azamfar, M.; Singh, J.; Bravo-Imaz, I.; Lee, J. Multisensor Data Fusion for Gearbox Fault Diagnosis Using 2-D Convolutional Neural Network and Motor Current Signature Analysis. *Mech. Syst. Signal Process.* **2020**, *144*, 106861. [CrossRef]
20. Li, F.; Pang, X.; Yang, Z. Motor Current Signal Analysis Using Deep Neural Networks for Planetary Gear Fault Diagnosis. *Meas. J. Int. Meas. Confed.* **2019**, *145*, 45–54. [CrossRef]
21. Han, T.; Yang, B.S.; Choi, W.H.; Kim, J.S. Fault Diagnosis System of Induction Motors Based on Neural Network and Genetic Algorithm Using Stator Current Signals. *Int. J. Rotating Mach.* **2006**, *2006*. [CrossRef]
22. Batko, W.; Pawlik, P.; Cioch, W.; Dąbrowski, D. Method and System for Monitoring and Diagnosis of Gear, Especially Gear Wheel Drive Bucket Wheel Excavator. Patent PL 225381 B1 2013, 28 April 2017.
23. Combet, F.; Zimroz, R. A New Method for the Estimation of the Instantaneous Speed Relative Fluctuation in a Vibration Signal Based on the Short Time Scale Transform. *Mech. Syst. Signal Process.* **2009**, *23*, 1382–1397. [CrossRef]
24. National Instruments Corporation. *LabVIEW, Order Analysis Toolkit User Manual*; National Instruments Corporation: Austin, TX, USA, 2005.
25. Pawlik, P. Single-Number Statistical Parameters in the Assessment of the Technical Condition of Machines Operating under Variable Load. *Eksploat. Niezawodn. Maint. Reliab.* **2019**, *21*, 164–169. [CrossRef]
26. Lees, A.W. Misalignment in Rigidly Coupled Rotors. *J. Sound Vib.* **2007**, *305*, 261–271. [CrossRef]
27. Jesse, S.; Hines, J.W.; Kuropatwinski, J.; Edmondson, A.; Carley, T.G. *Motor Shaft Misalignment Versus Efficiency Analysis*; The University of Tennessee: Knoxville, TN, USA, 1970.
28. Qi, X.; Yuan, Z.; Han, X. Diagnosis of Misalignment Faults by Tacholess Order Tracking Analysis and RBF Networks. *Neurocomputing* **2015**, *169*, 439–448. [CrossRef]
29. Kłosowski, G.; Rymarczyk, T.; Cieplak, T.; Niderla, K.; Skowron, Ł. Quality Assessment of the Neural Algorithms on the Example of EIT-UST Hybrid Tomography. *Sensors* **2020**, *20*, 3324. [CrossRef]
30. Kłosowski, G.; Rymarczyk, T.; Kania, K.; Świć, A.; Cieplak, T. Maintenance of Industrial Reactors Supported by Deep Learning Driven Ultrasound Tomography. *Eksploat. Niezawodn. Maint. Reliab.* **2020**, *22*, 138–147. [CrossRef]
31. Jeannette, L. *Introduction To Neural Networks: Design, Theory, and Applications*; California Scientific Software: Oak Grove, CA, USA, 1994.
32. Bishop, C. *Pattern Recognition and Machine Learning*; Springer: New York, NY, USA, 2006.
33. Naitzat, G.; Zhitnikov, A.; Lim, L.-H. Topology of Deep Neural Networks. *J. Mach. Learn. Res.* **2020**, *21*, 1–40.
34. Shang, Y. Resilient Group Consensus in Heterogeneously Robust Networks with Hybrid Dynamics. *Math. Methods Appl. Sci.* **2021**, *44*, 1456–1469. [CrossRef]
35. Krizhevsky, A.; Sutskever, I.; Hinton, G.E. ImageNet Classification with Deep Convolutional Neural Networks. *Commun. ACM* **2017**, *60*, 84–90. [CrossRef]

36. Srivastava, N.; Hinton, G.; Krizhevsky, A.; Salakhutdinov, R. Dropout: A Simple Way to Prevent Neural Networks from Overfitting. *J. Mach. Learn. Res.* **2014**, *15*, 1929–1958. [CrossRef]
37. Shang, Y. Couple-Group Consensus of Continuous-Time Multi-Agent Systems under Markovian Switching Topologies. *J. Frankl. Inst.* **2015**, *352*, 4826–4844. [CrossRef]
38. Okada, K. Negative Estimate of Variance-Accounted-for Effect Size: How Often It Is Obtained, and What Happens If It Is Treated as Zero. *Behav. Res. Methods* **2017**, *49*, 979–987. [CrossRef] [PubMed]
39. Cohen, J. *Statistical Power Analysis for the Behavioral Sciences*, 2nd ed.; Lawrence Erlbaum Associates: New York, NY, USA, 1988.

 energies

Article

Classification Trees in the Assessment of the Road–Railway Accidents Mortality

Edward Kozłowski [1], Anna Borucka [2,*], Andrzej Świderski [3] and Przemysław Skoczyński [3]

1. Faculty of Management, Lublin University of Technology, Nadbystrzycka 38, 20-618 Lublin, Poland; e.kozlovski@pollub.pl
2. Faculty of Security, Logistics and Management, Military University of Technology, ul. gen. Sylwestra Kaliskiego 2, 00-908 Warsaw, Poland
3. Motor Transport Institute, ul. Jagiellońska 80, 03-301 Warsaw, Poland; andrzej.swiderski@its.waw.pl (A.Ś.); przemyslaw.skoczynski@its.waw.pl (P.S.)

* Correspondence: anna.borucka@wat.edu.pl

Abstract: A special element of road safety research is accidents at the interface of the road and rail system. Due to their low share in the total number of incidents, they are not a popular subject of analyses but rather an element of collective studies, whereas the specificity of the road–rail accidents requires a separate characteristic, allowing, on the one hand, to categorize these types of incidents, and on the other, to specify the factors that affect them, along with an assessment of the strength of this impact. It is important to include in such analyses all potential predictors, both qualitative and quantitative. Moreover, the literature considers most often a number of accidents while, according to the authors, it does not fully reflect the scale of the danger. A better evaluation would be the victim's degree of injury. Therefore, the purpose of this article is to assess the likelihood of occurrence of various effects of road–rail accidents in the aspect of selected factors. Due to the ordinal form of the dependent variable, the classification trees method was used. The results obtained not only allow the characterization and assessment of the danger but also constitute guidelines for taking preventive actions.

Keywords: road–railway accidents; classification trees; road safety; transport means; accidents victims

Citation: Kozłowski, E.; Borucka, A.; Świderski, A.; Skoczyński, P. Classification Trees in the Assessment of the Road–Railway Accidents Mortality. *Energies* **2021**, *14*, 3462. https://doi.org/10.3390/en14123462

Academic Editor: Grzegorz Peruń

Received: 11 May 2021
Accepted: 9 June 2021
Published: 11 June 2021

1. Introduction

Railroad transport safety is an important factor taken into account when evaluating the operation of this branch of transport. Due to the importance, scope and consequences for society and economy of the low level of traffic safety, it is the subject of many studies and analyses and is systematically evaluated [1–3]. According to the latest annual reports/statistics from the International Union of Railways (UIC), the number of railroad accidents is decreasing [4,5]. The European Union Agency for Railways (ERA) reports are similarly optimistic. The latest reported safety level is historically the highest, although ERA points out that while safety levels have steadily improved, the rate of improvement has slowed down [6].

World statistics are derived from research, and thus, the presented trends are very important for decision-making also at the national level. It should be noted that the level of safety should be shaped primarily at the local level. In Poland, according to the latest data [7], both the overall number of railroad accidents and fatalities decreased. However, the rates of accidents at railroad crossings have not decreased. For example, in 2019, 11 more people died there than the year before. This shows that despite the overall positive indicators (both at the global and national level), there are areas for safety improvement and analysis in a smaller scope, in this case concerning only road–railway crossings. This became the genesis of this publication. Additionally, the UIC report shows that at the international level, railway crossing accidents account for as much as 15% of all diagnosed causes of railroad accidents, being the second most common cause (after persons trespassing on railroad infrastructure) estimated at 75% [5]. This further supports the analysis presented in the article. Another important argument is that, according to the ERA report [6], the overall level of safety at railroad crossings in Europe has improved. The average annual decrease in accidents between 2010 and 2018 was 3%, and 4% for fatalities. This shows that Poland does not fit into the general trend in Europe in this regard. Therefore, detailed analyses for the country are necessary.

Rail–road level crossings are an important element of the road infrastructure, enabling an intersection of the road and rail vehicles' tracks. There are over 14,000 such intersections in Poland [8] at which, in the period under investigation (the years 2014–2020), almost 8000 incidents (collisions and accidents) were recorded. Approximately 259 people were killed, and 404 were injured. Such accidents, apart from the tragic consequences, are associated with high costs, especially regarding the repair of rail vehicles. That is why the research is important, aimed to increase the level of safety and reduce the scale of the danger.

The accidents at the meeting point of road and rail transport are not a popular subject of analyses because they are relatively rare. This is shown in Figure 1, which shows the number of total road accidents during the study period and the number of road–railway accidents. The differences are so large that two scales had to be used to make the figure visible.

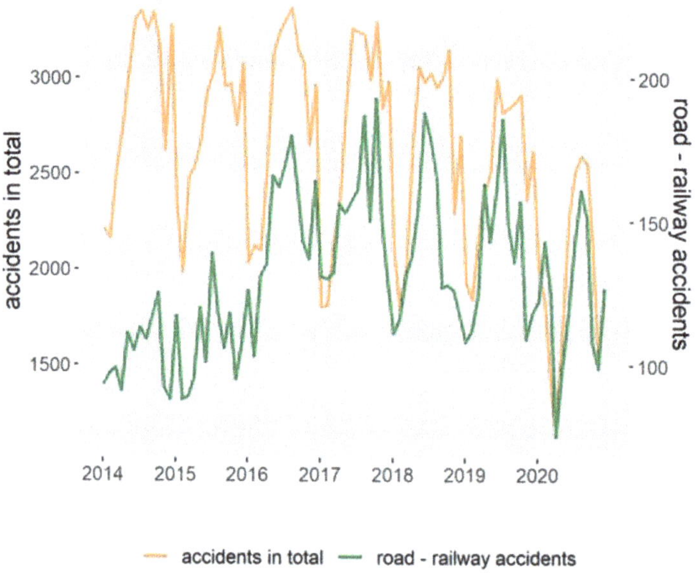

Figure 1. Number of total traffic accidents and road–railway accidents during the studied period.

Calculations show that road–rail accidents account for only 5% of all traffic incidents. Therefore, the interest in this issue is mainly due to the potential severity of their consequences. The research in the literature concerns, for example, predicting the likelihood of accidents, injuries and fatalities using logistic regression [9]. Ghomi and others in [10] used the ordered probit model, CART and association rules to evaluate the factors that most strongly affect the risk of an accident, which turned out to be train speed, age and gender of the incident participant.

The large-scale study was conducted in [11] and in [12]. The first concerns the analysis of the railroad level crossings in Great Britain over 64 years (1946–2009), while the second one concerns Australia (Victoria) in the years 1969–1974. However, the mentioned research is not new. The dynamic development of motorization, the constantly increasing number of road users, as well as the development of road safety systems [13] make it necessary to update them. This conclusion indicates that new analyses are needed in this area.

The accessibility of information does not help such analyses, which in many countries is very limited [14,15], which is why the data from simulators are used [14].

The most frequently studied element are factors affecting the number of accidents or their consequences. They differ from one country to another. In Israel [14], the warning device category, vehicle traffic intensity, train traffic intensity, and visibility conditions were considered important. In Ethiopia [16], the studies have shown that most accidents are due to human error, followed by technical problems and non-compliance with operational procedures. In [17], Ling et al. evaluated the derailments of Australian passenger trains as a result of a collision with heavy road trucks.

Classification models are popular both for the analysis of railroad accidents and all accidents in general [18,19]. The decision to use them is primarily influenced by the form of the dependent variable, as well as the nature of the factors that may affect the number of such incidents or the injuries caused by them. These are often qualitative variables, which limits the availability of some methods of mathematical analysis.

Most commonly used are logistic regression and decision trees, as the most popular tools for assessing the impact of qualitative predictors [20–24]. Decision trees were used, for example, in the investigation of car accidents from the years 2005 to 2006 in Taiwan, where the key factors determining the effects of injuries turned out to be driving under the influence of alcohol, not using seat belts, type of vehicle, type of collision, number of vehicles involved in the accident and location of the accident [25].

The authors [24] also used such a method to analyze 4-year observations regarding road accidents in India, indicating that in order to improve safety on motorways, first of all, it is necessary to properly design and control motorway entrances and limit the speed achieved by vehicles. The CART algorithm was used to investigate accidents in 2001 in Taiwan. This made it possible to evaluate the relationship between the severity of injuries and driver characteristics, vehicle type and conditions during an accident. The results indicate that the most important variable related to the severity of the collision is the vehicle type. It has also been identified that pedestrians, motorcyclists and cyclists run a greater risk of injury than other drivers in road accidents [26].

The logistic regression, in turn, was used in the study of falling asleep at the wheel or fatigue as factors contributing to an accident [27]. In [20], the probability of death was examined, indicating a significant impact of the location and cause of the accident. On the other hand, the authors [28] used polynomial logistic regression to identify significant factors of risk of accidents, indicating elements related to the road infrastructure, driver's characteristics and vehicle type.

Guided by the experiences resulting from reviewing the literature in terms of the tools used and the research gap identified regarding the railroad accidents analysis, the authors decided to use classification trees to evaluate factors influencing the effects of such an incident. Additionally, other approaches like random forest [29,30] and boosting technique [31–33] were used for improving the results obtained from the decision tree. All research was performed in the R environment.

2. Data for the Study

In Poland, information on traffic incidents is collected by the Police in the Accident and Collision Records System (SEWiK). Based on this, the Polish Road Safety Observatory (POBR), operating at the Motor Transport Institute (ITS), develops databases that have become the genesis of this study. The files made available contain a number of detailed characteristics for each incident, including:

- Place and time of the incident,
- Type of incident: collision of vehicles, rear-end collision, overturning,

- Consequences: fatalities, injured people (including slightly and seriously injured),
- Behavior of participants (drivers, passengers, pedestrians),
- Vehicle and road infrastructure condition,
- Atmospheric conditions (sun, rain, fog, strong wind), time of day and lighting (day, night, darkness, dawn),
- Type and condition of the road (surface, signposting, traffic lights),
- Circumstances and causes of the incident.

This study uses factors related to:

- Date and time of the incident,
- Existence of traffic lights at a railroad level crossing,
- Geographical location of the accident (province),
- Type of area (built-up, undeveloped),
- Driver characteristics (age, driving under the influence of alcohol or other drugs),
- Type of the injured participant (pedestrian, driver),
- Type of vehicle involved in the incident (passenger car, truck, truck with semi-trailer, motorcycle, motorcycle with an engine capacity of up to 125 cm^3, moped, bicycle, bus, agricultural tractor, train, emergency vehicle).

The data from 2014 to 2020 was examined. The total number of incidents per month, with the participation of rail vehicles in this period, is shown in Figure 1 (green line).

Almost 11,000 people took part in these accidents, most of whom have not been injured. Other victims, according to the type of injuries sustained, were divided into three categories [34]:

- Fatal—people who died at the scene of the accident or within 30 days of the date of the accident due to the injuries sustained,
- Seriously injured—persons suffered severe disability, a serious incurable illness or long-term, life-threatening illness, permanent mental illness, complete or significant permanent incapacity for work or permanent, significant deformation or deformity of the body; this term also includes a person who suffered other injuries resulting in violation of bodily functions or health disorder for a period of more than 7 days,
- Slightly injured—injuries other than listed above and causing a health disorder in the period of no more than 7 days.

Due to the fact that almost 93% of participants in incidents did not suffer any injuries during the period under investigation, it was decided to further analyze only those who suffered health damage or died.

Results of the analysis of road traffic safety in Poland indicate that in the studied period there were 8474 accidents and collisions. Participants in them were 10,960 people, of which 210 died. This is less than 2% of all victims. For comparison, the total number of all road accidents and collisions in this period was 219,863. As many as 18,527 people died in them. Reducing the number of people injured in road accidents is the main goal of all actions undertaken in the area of road safety improvement. For this reason, the authors found it necessary to focus only on accidents and their worst effects. This will allow identifying the most important causes (factors) of these accidents and thus the necessary preventive actions (making the right decisions).

3. Decision Trees

We consider the feature Y (called the injured state), which depends on the value of the features (independent variables) $X_1, \ldots, X_m$ presented in the previous chapter. One of the possible ways to determine the relationship between features is to construct a decision or regression tree. In the presented case, the Y feature is qualitative, so in order to analyze the impact of incident circumstances on the injured person's condition, decision trees have been used.

Let $D = \left\{ \left(x_{(i)}, y_i \right) : x_{(i)} \in R_1 \times \ldots \times R_m \,,\, y_i \in A,\, 1 \leq i \leq n \right\}$ be the learning set. For any $1 \leq j \leq m$ set R_j denotes the possible realization of X_j feature, but set A denotes the set of possible realizations of a response variable, where cardinality $\#A = h > 0$ (power of the A set or the number of possible classes is equal h). For the $i-$th observation, $1 \leq i \leq n$ the vector $x_{(i)} \in R_1 \times \ldots \times R_m$ denotes the realizations of independent (input) variables (usually feature values that influence to output variable) but $y_i \in A$ denotes the value of the response variable. Our task is to define a model, where based on observations $x \in R_1 \times \ldots \times R_m$ we should predict a victim condition. To assess the features influence on the sufferer state, we will apply the decision tree model.

The tree-based method consists of partition (splitting, division) of the feature space $S = R_1 \times \ldots \times R_m$ into a set of separable regions and fitting values of a response variable to appropriate regions. Below, we consider a decision problem for response variable Y. We split the entire S feature space into $S_1, S_2, \ldots, S_k$ regions, where $S_i \cap S_j = \varnothing$ for $1 \leq i \neq j \leq k$. Based on input vector $x \in S$, we predict the output variable Y as follows:

$$f(x) = \sum_{j=1}^{k} c_j I_j(x), \tag{1}$$

where,

$$I_j(x) = \begin{cases} 1, & for\ x \in S_j \\ 0, & for\ x \notin S_j \end{cases} \tag{2}$$

and value $c_j \in A$ for $1 \leq j \leq k$ denotes the most commonly occurring class of response variable in the S_j region. From (1), we see that the main task during decision tree building consists of splitting the entire space of features into separated regions.

The regression tree is usually presented in graphic form. The internal tree nodes describe how the division was made, while the leaves correspond to the classes to which the objects belong. The tree edges, in turn, represent the values of the features based on which the division was made.

For each S_j region we estimate the classification rates $0 \leq p_{j1}, \ldots, p_{jh}$ corresponding to elements from set A (possible realizations of response variable) where $p_{j1} + \ldots + p_{jh} = 1$. The value p_{ji} represents the proportion of observations in the $j-$th region that are from the $i-$th class. The classification error rate is a fraction of observations in this region that do not belong to the most common class

$$error_j = 1 - \max_{1 \leq i \leq h} p_{ji} \tag{3}$$

The decision tree building method consists of portioning of an appropriate region by minimizing the Gini index

$$G_j = \sum_{i=1}^{h} p_{ji}(1 - p_{ji}) \tag{4}$$

or entropy

$$E_j = -\sum_{i=1}^{h} p_{ji} \log p_{ji} \tag{5}$$

From (4) and (5) we can see that the Gini index and entropy take on a small value when the classification rates $p_{j1}, \ldots, p_{jh}$ are close to zero or one. Both the Gini index and entropy are referred to as purity of $j-$th node and typically used to assess the quality of a particular split of a region.

The restrictions that can be applied during the division of S feature space are: the minimum cardinality of node subject to dividing, the minimum cardinality of the node resulting from dividing, the maximum number of tree levels. Selecting the right tree size can also be adjusted by pruning the original model.

For this purpose, there are selected algorithms being used. Among the most popular are: CART and C4.5 (and then C5.0) algorithms. Additionally, CHAID [35], QUEST, THAID and others [36] can be used. In our analysis, we employ the CART algorithm to select the important features' influence on accidents result.

The decision trees suffer from high variance. One of the possible techniques to improve the predictions obtained from decision trees is bagging [33]. The main idea depends on creating an ensemble of decision trees based on several bootstrapped training sets. These training sets are chosen randomly with replacement from the data set and are used to train the decision trees. The variance is reduced by aggregating a set of predictions obtained from an ensemble of trees. For classification trees, we take a majority vote from the obtained class predicted by each tree.

The random forest is an extension of the bagging method [29,30]. The main difference is that during the making of the training set for each tree we randomly choose the set of features from a full set of features. Thus, we make an ensemble of random trees. A multitude of random trees is called a random forest. This technique avoids the problem of selecting the dominant predictor in the split of space for each tree. The predictions obtained from trees with randomly selected features are less correlated, thereby making the average of the predictions obtained from regression trees or majority vote from classification trees less variable and more reliable.

Another technique to improve the results obtained from the decision trees is boosting [33]. Boosting, like bagging, depends on creating an ensemble of decision trees. For bagging, we adapt the trees to training sets chosen randomly from the data set. By applying the boosting technique, the trees are made sequentially, i.e., the current tree is built based on information from the previously grown tree and the response variable in the current tree is defined as residuals (not explained outcomes) from the previous tree.

The adaptation of a large decision tree to the data can be hard and potentially overfitting. The boosting approach results in the learning process being slow. By adding the current decision tree into the ensemble of trees in order to update the residuals, we define the model that explains the dependences between outcomes and features. Each of these trees can be small but by adapting the small trees to the residuals, we improve the outcomes (response variable) in areas where this does not work well. It is the main benefit of this method. In general, the learning process is slow and sequential but tends to explain the dependences well. In our analysis, we employ the XGB (eXtreme Gradient Boosting) algorithm [31,32] to select the influence of important features on the accidents' result.

Various measures are used to evaluate the classifier. Most often, the basis for their definition is the confusion matrix. The columns of this matrix determine the actual decision classes while rows determine the decisions predicted by the model. The N_{ij} value at the intersection of the $i-$th verse and $j-$th column specifies the number of observations of $j-$th class classified into the $i-$th class, $1 \leq i,j \leq h$. In general, the case has the form presented in Table 1.

Table 1. Form of the confusion matrix.

Actual Class → **Predicted Class ↓**	**Class 1**	**Class 2**	**...**	**Class *h***
Class 1	N_{11}	N_{12}	...	N_{1h}
Class 2	N_{21}	N_{22}	...	N_{2h}
...	...		...	
Class *h*	N_{h1}	N_{h2}	...	N_{hh}

For each possible realization, we estimate the basic values. For the $j-$th class ($1 \leq j \leq h$), the TP (true positive) denotes a number of outcomes (instances) that are correctly classified for this class, the FP (false positive) is the number of outcomes that are classified for the class but they do not belong to it, the FN (false negative) is the number of outcomes that belong to the class but are incorrectly classified, the TN (true negative) is the number of correctly classified outcomes that do not belong to the class. According to the notation presented in Table 1 for the j−th class ($1 \leq j \leq h$), we determine the basic values as follows:

$$TP = N_{jj},\ FP = \sum_{\substack{i=1,\\ i\neq j}}^{h} N_{ji}, \tag{6}$$

$$FN = \sum_{\substack{i=1,\\ i\neq j}}^{h} N_{ij},\ TN = \sum_{i,j=1}^{h} N_{ij} - TP - FP - FN \tag{7}$$

Additionally, for each class, we estimate the following basic metrics:

1. Sensitivity (Recall, True Positive Rate—***TPR***), indicating to what extent the truly positive class has been classified as positive:

$$\boldsymbol{TPR} = \frac{TP}{TP + FN} \tag{8}$$

2. Specificity (True-Negative Rate—***TRN***), indicating to what extent the truly negative class has been classified as negative:

$$\boldsymbol{TNR} = \frac{TN}{TN + FP} \tag{9}$$

3. Positive Predictive Value (***PPV***), indicating with what certainty we can trust positive predictions, i.e., in what percentage are the positive predictions confirmed by the truly positive state:

$$\boldsymbol{PPV} = \frac{TP}{TP + FP} \tag{10}$$

4. Negative Predictive Value (***NPV***), indicating with what certainty can we trust negative predictions, i.e., in what percentage the negative predictions are confirmed by the truly negative state:

$$\boldsymbol{NPV} = \frac{TN}{TN + FN} \tag{11}$$

5. ***Prevalence*** is the fraction of cases possessing the examined feature (it shows how often the positive class occurs in the sample).

$$\boldsymbol{Prevalence} = \frac{TP + FN}{TP + TN + FP + FN} \tag{12}$$

6. ***Detection rate*** shows the number of correct positive class predictions as a proportion of all of the predictions made.

$$\boldsymbol{Detection\ rate} = \frac{TP}{TP + TN + FP + FN} \tag{13}$$

7. ***Detection prevalence*** or predicted positive condition rate (PPCR) is the percentage of observations that the classifier predicted as positive (it illustrates the feasibility of the model in practice).

$$\boldsymbol{Detection\ Prevalence} = \frac{TP + FP}{TP + TN + FP + FN} \tag{14}$$

8. ***Balanced accuracy*** is an average arithmetic sensitivity and specificity, specifying the average number of predictions for each class, correctly classified by the model (it finds better use when we have just one test set, and it is not balanced).

$$\boldsymbol{Ballanced\ Accuracy} = \frac{TPR + TNR}{2} \tag{15}$$

Additionally, for the entire classifier, we determine accuracy (*ACC*), which denotes the fraction of all instances that are correctly categorized:

$$ACC = \frac{\sum_{i=1}^{h} N_{ii}}{\sum_{i,j=1}^{h} N_{ij}} \quad (16)$$

4. Results

The CART algorithm was used for the construction of decision trees. The influence of the characteristics of the province and the time of the incident on the condition of the victim was investigated. There were 631 observations used for the construction. The following symbols were adopted for individual provinces: B—Podlaskie, C—Kujawsko-Pomorskie, D—Dolnośląskie, E—Łódzkie, F—Lubuskie, G—Pomorskie, K—Małopolskie, L—Lubelskie, N—Warmińsko-Mazurskie, O—Opolskie, P—Wielkopolskie, R—Podkarpackie, S—Śląskie, T—Świętokrzyskie, W—Mazowieckie, Z—Zachodnio-Pomorskie. Figure 2 presents the classification tree with maximum depth equalling 5. This tree contains only seven rules.

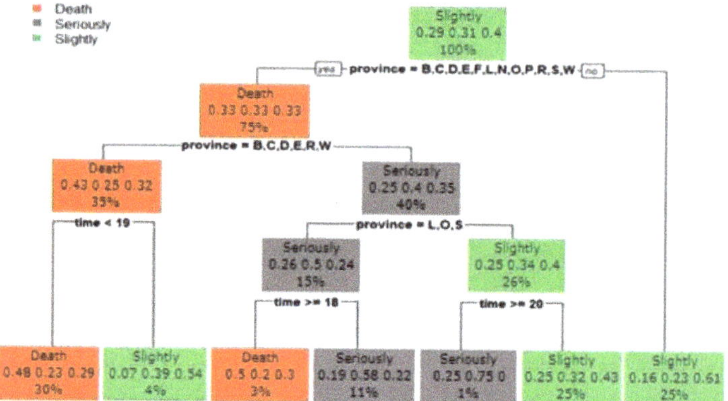

Figure 2. Classification tree for the variables of place and time of the accident.

The accuracy of the model taking into account only two variables is not satisfactory, with the value of ACC = 0.517. The accuracy of the predictions is presented by the confusion matrix—Table 2. The elements on the main diagonal indicate correctly classified observations.

Table 2. Confusion matrix for the decision tree based on location of incident and time.

	Class		
Prediction	**Death**	**Seriously**	**Slightly**
Death	102	49	61
Seriously	16	48	16
Slightly	66	97	176

Because the quality of the prediction is not satisfactory, a decision tree was constructed that includes a higher number of predictors. The following variables were taken into account: time of day, month, type of vehicle, type of participant, existence of traffic lights at the level crossing, age of the victim, area (developed, undeveloped), and location of the incident (province).

The classifier includes 49 decision rules. Its extensive form prevents legible, graphical presentation. Therefore, only a descriptive characteristic of the model was made using a matrix of errors, basic measures of states and a graph with variable importance.

The inclusion of additional predictors has improved the quality of the classifier. The accuracy is ACC = 0.679. Table 3 shows the confusion matrix. On the main diagonal, there are correctly classified observations.

Table 3. Confusion matrix for the extended decision tree.

	Class		
Prediction	**Death**	**Serious Injury**	**Slight Injury**
Death	149	60	50
Seriously	13	95	18
Slightly	22	39	185

The predictors, ranked by their importance, are shown in Figure 3. This figure shows the percentage of decrease of the Gini index (interpreted as gain) for the construction of the classification tree. The evaluation of the importance of the predictors' impact on the dependent variable has indicated a significant impact primarily of the location of the accident (province) and the time of the incident. A detailed evaluation of the model was made by analyzing measures for each of the singled-out injury levels. The sensitivity and specificity take values exceeding 73% (except for the "serious injury" class for which sensitivity is about 49%). The precision of positive and negative prediction is high (up to 75% except for the "death" class). The remaining results are presented in Table 4.

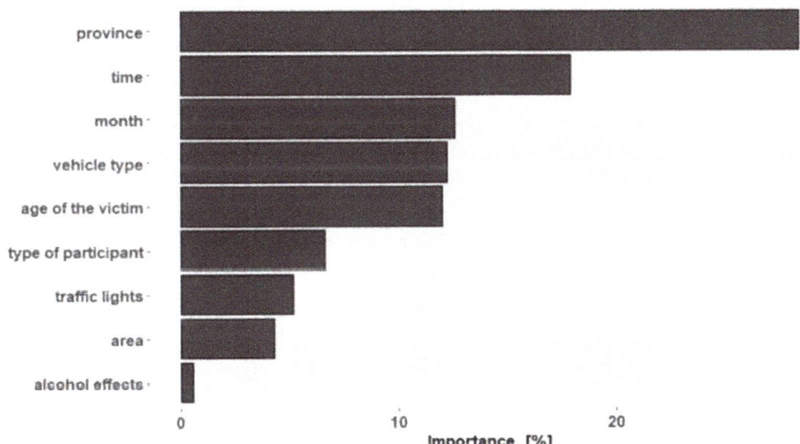

Figure 3. Variable importance ranking for the extended decision tree.

Table 4. Detailed measures for the individual injury classes.

Class	Death	Serious Injury	Slight Injury
Sensitivity	0.8098	0.4897	0.7312
Specificity	0.7539	0.9291	0.8386
Pos Pred Value	0.5753	0.7540	0.7520
Neg Pred Value	0.9059	0.8040	0.8234
Prevalence	0.2916	0.3074	0.4010
Detection Rate	0.2361	0.1506	0.2932
Detection Prevalence	0.4105	0.1997	0.3899
Balanced Accuracy	0.7818	0.7094	0.7849

In order to verify the influence of each feature, a random forest was also constructed, consisting of 50 trees, where three features were randomly selected for the learning set. As before, we present the results in the form of a confusion matrix (Table 5), a matrix of the basic measures for each state (Table 6) and the variable importance plot (Figure 4).

Table 5. Confusion matrix for the random forest.

		Class	
Prediction	**Death**	**Serious Injury**	**Slight Injury**
Death	175	16	8
Seriously	3	160	4
Slightly	6	18	241

Table 6. Detailed measures of the random forest for the individual injury classes.

Class	Death	Serious Injury	Slight Injury
Sensitivity	0.9511	0.8247	0.9526
Specificity	0.9463	0.9840	0.9365
Pos Pred Value	0.8794	0.9581	0.9094
Neg Pred Value	0.9792	0.9267	0.9672
Prevalence	0.2916	0.3074	0.4010
Detection Rate	0.2773	0.2536	0.3819
Detection Prevalence	0.3154	0.2647	0.4200
Balanced Accuracy	0.9487	0.9044	0.9445

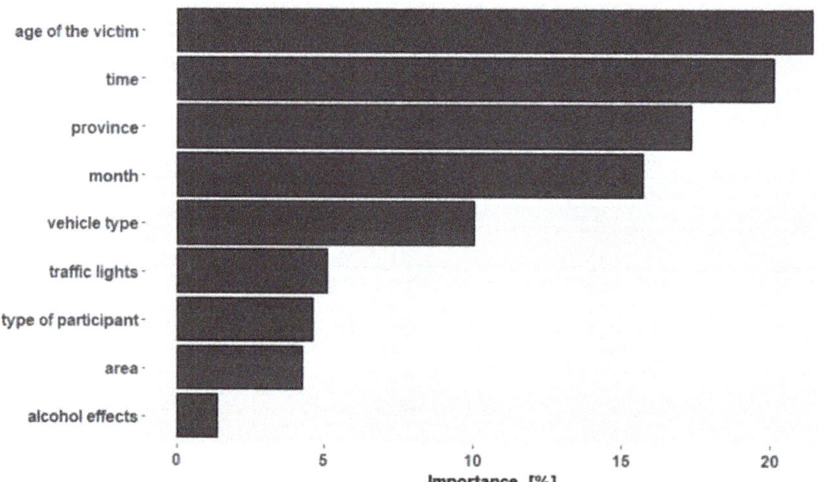

Figure 4. Predictor importance ranking for the random forest.

By comparing Tables 3 and 5, we can see that the quality of the classifier for the random forest is better than for the extended decision tree. For the random forest, the accuracy is equal to 0.913.

From Figure 4, we can see that the predictors: age of the victim, time, province and month have the greatest impact on the variable describing the state of the injured person. Basic metrics for the random forest are presented in Table 6.

From Table 6, we see a significant improvement in metrics sensitivity, specificity, balanced accuracy, PPV and NPV.

The classifier was also constructed using the boosting technique. In this case, it is assumed that the maximum depth of a tree equals 3, and the maximum number of boosting iterations is 50. As before, we present the results in the form of a confusion matrix (Table 7), matrix of the basic measures for each state (Table 8) and a variable importance plot (Figure 5).

Table 7. Confusion matrix for the boosting tree model.

		Class	
Prediction	**Death**	**Serious Injury**	**Slight Injury**
Death	179	5	1
Seriously	3	185	3
Slightly	2	4	249

Table 8. Detailed measures of boosting tree model.

Class	Death	Serious Injury	Slight Injury
Sensitivity	0.9728	0.9536	0.9842
Specificity	0.9866	0.9863	0.9841
Pos Pred Value	0.9676	0.9686	0.9765
Neg Pred Value	0.9888	0.9795	0.9894
Prevalence	0.2916	0.3074	0.4010
Detection Rate	0.2837	0.2932	0.3946
Detection Prevalence	0.2932	0.3027	0.4041
Balanced Accuracy	0.9797	0.9699	0.9842

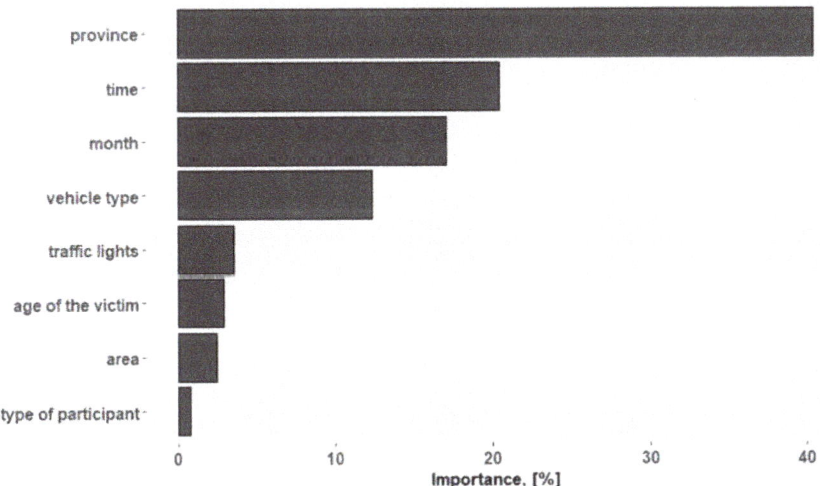

Figure 5. Predictor importance ranking for the boosting tree model.

By comparing Tables 3, 5 and 7, we can see that the quality of the classifier for the boosting tree model is the best. For this model, the accuracy equals 0.972.

From Figure 5, we can see that the province predictor dominates the others. The time, month and vehicle type predictors have a significant impact on the variable describing the state of the injured person. Basic metrics for the boosting tree model are presented in Table 8.

From Table 8, we see that the sensitivity, specificity, balanced accuracy, PPV and NPV metrics are over 0.95.

The most powerful predictor, in the case of decision trees and the boosting tree model, is the location of the accident, in this case defined by the province. This can be due to two reasons. The first is the discrepancy in the state and quality of both linear and point infrastructure in the individual regions of Poland. These elements remain in the sphere of administration of the local government units. This condition applies to as much as 95.4% of all roads in the country and means different management, financing and control. Because the impact of infrastructure on both the number of accidents and injuries is significant [37], the impact of location in the presented study is also significant. Another reason influencing this result is the different densities of the railroad network in individual regions of Poland. In eastern Poland, it is smaller than in western Poland. However, central Poland has the largest number of railroad lines. Moreover, the number of road users as well as their mobility vary between provinces. This activity, different not only in particular areas but also during the times of the day, results in the next factor strongly influencing the model being the time of the event.

Another important predictor related to time is the month. The increased number of traffic accidents is probably influenced by the increase in traffic associated with vacation activity from May to September. It may also be due to the reduced alertness of drivers focused on resting. The fact that some drivers are not daily users of cars and have little experience in driving also contributes to the accidents. The fourth most important is the type of vehicle. Among the established rules, the most common means of transport is the bicycle, the passenger car and the truck. Additionally, in many situations, cyclists' deaths are equal to or very close to 100%.

In the case of the random forest, the age of the victim was the first factor contributing to injury in accidents. The other factors with the highest influence were the same but ranked in a different order (time, province, month).

The influence of the remaining predictors was significant but smaller. The existence of traffic lights and developed areas is conducive to the tragic consequences of accidents. The drivers are more likely to die than passengers. The influence of alcohol is also an important factor, but it should be emphasized that this result is affected by a small number of accidents in which such a violation was noted. In the analyzed sample, there is less than 1.2% of them.

5. Conclusions

The classification trees are a flexible, user-friendly, and easy-to-interpret tool for analyzing large sets of observations, consisting of many variables. Their biggest advantages include transparency and readability of the result presented in the form of rules, as well as no requirements for the form and distribution of variables. Additionally, it is necessary to emphasize their insensitivity to the occurrence of non-typical observations and deficiencies in the data set.

However, in the analyzed case, the application of this method resulted in an accuracy of 68%. This result was improved by applying the boosting tree model for which the highest accuracy of 97% was achieved. In both cases, the result was the same. The most important predictors were: province, time, month and vehicle time. An additional method proposed was random forest, for which ACC = 91%. This model indicated a different order of predictors, placing the age of the victim first.

The results obtained, in addition to indicating the factors increasing the risk of certain injuries in an accident, also indicate the need to develop comprehensive solutions for the entire country in terms of improving road safety. Such a large impact of location may result from the different functioning of individual local government units and the differences in administration of the governed infrastructure. Therefore, it is necessary to develop and implement common standards and equalize the differences between individual regions, particularly in relation to the condition of the road, its surroundings and road equipment.

The obtained measurable results of measurements concerning the influence of the examined factors on traffic safety at railroad crossings provide information on such actions, which are conducive to improving road–rail traffic safety through:

- Shaping of the road–railway traffic safety strategy based not only on the analysis of data on the number of road–railway incidents but also on the factors influencing mortality and the strength of this influence,
- Ensuring funds in the state budget for the creation and improvement of national and local databases collecting information not only on the number of road–railway accidents but also covering detailed characteristics of each event (e.g., visibility range at a railroad crossing, number of lines/tracks–track gauge, frequency of railroad links),
- Setting standards for improving the safety of road–railway infrastructure in terms of traffic engineering and road and construction issues,
- Setting standards for improving the safety of road–railway infrastructure for owners and managers of roads with road–railway connections on each administrative level,
- Shaping the behavior of all road users and awareness of existing risk factors and the significance of their impact on road–railway incidents and their consequences,
- Conducting social campaigns shaping attitudes and opinions, also on the basis of obtained research results showing which factors most strongly influence mortality in road–railway accidents,
- Creating and enforcing stricter regulations, especially with respect to identified causes of fatal accidents, and increasing the penalties in this area,
- Improving the operation of road–railway rescue systems by identifying areas (railroad crossings) particularly conducive to fatal accidents,
- Improving the process of education and training using the results of analysis of factors affecting the mortality in road–railway accidents for prevention purposes, as part of training and prevention talks, as well as guidelines for determining the timing and scope of police operations organized in support of safety.

A limitation of the analysis presented in this paper is, first of all, the qualitative form of most of the variables. It limits the possibility of research to classificatory methods. The quality of the presented research is also affected by the number of recorded factors. Especially in the framework of further considerations, the authors would like to take into account the volume of traffic. This factor is very important from the point of view of road–railway traffic safety, but it is not monitored at most of the railroad crossings. Generally speaking, traffic volume monitoring in Poland concerns mainly selected regions (mostly intersections of big cities). However, the dynamic development of smart transportation systems is conducive to obtaining the necessary information [38], so analyses that take this factor into account for a smaller area will probably be possible soon.

Author Contributions: Conceptualization, A.B., E.K. and A.Ś.; methodology, A.B. and E.K. software, A.B. and E.K.; validation, A.B. and E.K.; formal analysis, A.B. and E.K.; investigation, A.B. and E.K.; resources, A.Ś. and P.S.; data curation, A.Ś. and P.S.; writing—original draft preparation, A.B., E.K. and A.Ś.; writing—editing, A.B., E.K. and A.Ś.; visualization, A.B. and E.K. All authors have read and agreed to the published version of the manuscript.

Funding: This research received no external funding.

Institutional Review Board Statement: Not applicable.

Informed Consent Statement: Not applicable.

Data Availability Statement: The data presented in this study are openly available in a publicly accessible Mendeley Data repository: http://dx.doi.org/10.17632/7vwx2pnbx5.1 (accessed on 11 May 2021).

Conflicts of Interest: The authors declare no conflict of interest.

References

1. Blagojević, A.; Kasalica, S.; Stević, Ž.; Tričković, G.; Pavelkić, V. Evaluation of safety degree at railway crossings in order to achieve sustainable traffic management: A novel integrated fuzzy mcdm model. *Sustainability* **2021**, *13*, 832. [CrossRef]
2. Cao, T.; Mu, W.; Gou, J.; Peng, L. A study of risk relevance reasoning based on a context ontology of railway accidents. *Risk Anal.* **2020**, *40*, 1589–1611. [CrossRef] [PubMed]
3. Li, K.; Pan, Y. An effective method for identifying the key factors of railway accidents based on the network model. *Int. J. Mod. Phys. B* **2020**, *34*, 2050192. [CrossRef]
4. Batarlienė, N. Improving safety of transportation of dangerous goods by railway transport. *Infrastructures* **2020**, *5*, 54. [CrossRef]
5. EUAFR. *UIC Annual Report*; European Union Agency for Railways: Luxembourg, 2020.
6. EUAFR. *Report on Railway Safety and Interoperability in the EU-2020*; Publications Office of the European Union: Luxembourg, 2020.
7. UTK. *Raport w Sprawie Bezpieczeństwa Transportu Kolejowego w Polsce w 2019 r*; Dziennik Urzędowy Prezesa Urzędu Transportu Kolejowego nr 16/2020; UTK: Warsaw, Poland, 2020.
8. NIK. *Bezpieczeństwo Eksploatacji Pasażerskiego Taboru Kolejowego. Informacja o Wynikach Kontroli*; Departament Infrastruktury: Warsaw, Poland, 2021.
9. McCollister, G.M.; Pflaum, C.C. A model to predict the probability of highway rail crossing accidents. *Proc. Inst. Mech. Eng. Part F J. Rail Rapid Transit* **2007**, *2213*, 321–329. [CrossRef]
10. Ghomi, H.; Bagheri, M.; Fu, L.; Miranda-Moreno, L.F. Analyzing injury severity factors at highway railway grade crossing accidents involving vulnerable road users: A comparative study. *Traffic Inj. Prev.* **2016**, *178*, 833–841. [CrossRef]
11. Evans, A.W. Fatal accidents at railway level crossings in Great Britain 1946–2009. *Accid. Anal. Prev.* **2011**, *435*, 1837–1845. [CrossRef]
12. Wigglesworth, E.C. Human factors in level crossing accidents. *Accid. Anal. Prev.* **1978**, *103*, 229–240. [CrossRef]
13. Desai, A.; Singh, J.J.; Spicer, M.T. Intelligent transport system to improve safety at road-rail crossings. In Proceedings of the 11th World Level Crossing Symposium, Tokyo, Japan, 26–29 October 2010.
14. Gitelman, V.; Hakkert, A.S. 1997 The evaluation of road-rail crossing safety with limited accident statistics. *Accid. Anal. Prev.* **1997**, *292*, 171–179. [CrossRef]
15. Weatherford, B.A.; Willis, H.H.; Ortiz, D.S.; Mariano, L.T.; Froemel, J.E.; Daly, S.A. *The State of US Railroads: A Review of Capacity and Performance Data Rand Corporation*; RAND Corporation: Santa Monica, CA, USA, 2008.
16. Ugochukwu, A.L.; Lovejoy, M.; Mercy, A. Safety demonstration and risk management at rail-road level crossing at addis ababa light rail transit network. *IJSRSET* **2019**, *65*, 103–109. [CrossRef]
17. Ling, L.; Dhanasekar, M.; Thambiratnam, D.P. Assessment of road-rail crossing collision derailments on curved tracks. *Aust. J. Struct. Eng.* **2017**, *182*, 125–134. [CrossRef]
18. Huang, W.; Liu, Y.; Zhang, Y.; Zhang, R.; Xu, M.; De Dieu, G.J.; Shuai, B. Fault Tree and Fuzzy DS Evidential Reasoning combined approach: An application in railway dangerous goods transportation system accident analysis. *Inf. Sci.* **2020**, *520*, 117–129. [CrossRef]
19. Rungskunroch, P.; Jack, A.; Kaewunruen, S. Risk and resilience of railway infrastructure: An assessment on uncertainties of rail accidents to improve risk and resilience through long-term data analysis. In *Lecture Notes in Civil Engineering*; Springer Nature: Basingstoke, UK, 2021.
20. Al-Ghamdi, A.S. Using logistic regression to estimate the influence of accident factors on accident severity. *Accid. Anal. Prev.* **2002**, *346*, 729–741. [CrossRef]
21. Kozłowski, E.; Mazurkiewicz, D.; Żabiński, T.; Prucnal, S.; Sęp, J. Assessment model of cutting tool condition for real-time supervision system. *Eksploat. Niezawodn. Maint. Reliab.* **2019**, *21*, 679–685. [CrossRef]
22. Rungskunroch, P.; Jack, A.; Kaewunruen, S. Benchmarking on railway safety performance using Bayesian inference, decision tree and petri-net techniques based on long-term accidental data sets. *Reliab. Eng. Syst. Saf.* **2021**, *213*, 107684. [CrossRef]
23. Rymarczyk, T.; Kozłowski, E.; Kłosowski, G.; Niderla, K. Logistic regression for machine learning in process tomography. *Sensors* **2019**, *19*, 3400. [CrossRef]
24. Singh, G.; Sachdeva, S.N.; Pal, M. M5 model tree based predictive modeling of road accidents on non-urban sections of highways in India. *Accid. Anal. Prev.* **2016**, *96*, 108–117. [CrossRef]
25. Chang, L.Y.; Chien, J.T. Analysis of driver injury severity in truck-involved accidents using a non-parametric classification tree model. *Saf. Sci.* **2013**, *511*, 17–22. [CrossRef]
26. Chang, L.Y.; Wang, H.W. Analysis of traffic injury severity: An application of non-parametric classification tree techniques. *Accid. Anal. Prev.* **2006**, *385*, 1019–1027. [CrossRef]
27. Sagberg, F. Road accidents caused by drivers falling asleep. *Accid. Anal. Prev.* **1999**, *316*, 639–649. [CrossRef]
28. Yan, X.; Radwan, E.; Abdel-Aty, M. Characteristics of rear-end accidents at signalized intersections using multiple logistic regression models. *Accid. Anal. Prev.* **2005**, *376*, 983–995. [CrossRef]
29. Breiman, L. Bagging predictors. *Mach. Learn.* **1996**, *26*, 123–140. [CrossRef]
30. Breiman, L. Random forests. *Mach. Learn.* **2001**, *45*, 5–32. [CrossRef]
31. Biecek, P.; Burzykowski, T. *Explanatory Model Analysis. Explore, Exam and Examine Predictive Models*; Chapman & Hall Book: London, UK, 2020.

32. Chen, T.; Guestrin, C. XGBoost: A scalable tree boosting system. In Proceedings of the 22nd ACM SIGKDD International Conference on Knowledge Discovery and Data Mining, San Francisco, CA, USA, 13–17 August 2016; pp. 785–794.
33. Hastie, T.; Tibshirani, R.; Friedman, J. *The Elements of Statistical Learning*; Springer-Verlag New York Inc.: New York, NY, USA, 2009.
34. Choinski, K. *Zarządzenie Komendanta Głównego Policji z Dnia 30 Czerwca 2006 r*; Komenda Główna Policji: Warsaw, Poland, 2006.
35. Kass, G.V. An exploratory technique for investigating large quantities of categorical data. *J. R. Stat. Soc. Ser. C Appl. Stat.* **1980**, *29*, 119–127. [CrossRef]
36. Loh, W.Y.; Shih, Y.S. Split selection methods for classification trees. *Stat. Sin.* **1997**, *7*, 815–840.
37. Graczyk, B.; Polasik, R. Wpływ infrastruktury drogowej na bezpieczeństwo ruchu drogowego. *Postępy Inżynierii Mech.* **2016**, *7*, 5–15.
38. Jamal, A.; Mahmood, T.; Riaz, M.; Al-Ahmadi, H.M. GLM-based flexible monitoring methods: An application to real-time highway safety surveillance. *Symmetry* **2021**, *13*, 362. [CrossRef]

Article

Optimal Pricing of Vehicle-to-Grid Services Using Disaggregate Demand Models

Charilaos Latinopoulos *, Aruna Sivakumar and John W. Polak

Department of Civil and Environmental Engineering, Imperial College London, London SW7 2BU, UK; a.sivakumar@imperial.ac.uk (A.S.); j.polak@imperial.ac.uk (J.W.P.)
* Correspondence: charilaos.latinopoulos10@imperial.ac.uk

Abstract: The recent revolution in electric mobility is both crucial and promising in the coordinated effort to reduce global emissions and tackle climate change. However, mass electrification brings up new technical problems that need to be solved. The increasing penetration rates of electric vehicles will add an unprecedented energy load to existing power grids. The stability and the quality of power systems, especially on a local distribution level, will be compromised by multiple vehicles that are simultaneously connected to the grid. In this paper, the authors propose a choice-based pricing algorithm to indirectly control the charging and V2G activities of electric vehicles in non-residential facilities. Two metaheuristic approaches were applied to solve the optimization problem, and a comparative analysis was performed to evaluate their performance. The proposed algorithm would result in a significant revenue increase for the parking operator, and at the same time, it could alleviate the overloading of local distribution transformers and postpone heavy infrastructure investments.

Keywords: electric vehicle charging; vehicle-to-grid; genetic algorithms; particle swarm optimization; demand-side management; discrete choice theory; revenue management

Citation: Latinopoulos, C.; Sivakumar, A.; Polak, J.W. Optimal Pricing of Vehicle-to-Grid Services Using Disaggregate Demand Models. *Energies* **2021**, *14*, 1090. https://doi.org/10.3390/en14041090

Academic Editors: Albert Y.S. Lam and Javier Contreras

Received: 3 November 2020
Accepted: 12 February 2021
Published: 19 February 2021

1. Introduction

Advances in battery technology, the low emission factors, the low operation costs and the high fuel economy of Battery Electric Vehicles (BEVs) and Plugged-in Hybrid Electric Vehicles (PHEVs) are some of the reasons that the family of Electric Vehicles (EVs) has attracted a lot of attention over the last few years. New models with extended capabilities and longer electric ranges are presented every year by major automobile manufacturers [1]. The latest generation of EVs shifts the paradigm towards new markets by providing extended full electric ranges of over 300 km and significantly reducing the problems associated with "range anxiety" [2].

The adoption of EVs in private transportation could ultimately lead to a replacement of crude oil with cleaner energy sources. At the same time, they can be transformed from unidirectional devices that draw power from the grid to bidirectional assets that transfer power back. Vehicle-to-Grid (V2G) may enable drivers to provide ancillary services to the grid in exchange for financial returns, as well as to contribute in alleviating peak power demand [3]. Kempton and Tomic [3,4] have compared existing grid services (spinning frequency reserves, peak power supply and regulation) with Vehicle-to-Grid (V2G) support and concluded that using EVs for regulation can offer the most substantial returns to vehicle owners.

The accelerated growth in electric mobility also demands the development of new methods to address their implications for the power grid. The stability of the power system is at stake, particularly when charging events cluster in space and time. Without charging coordination, the variations in charging demand could have a great impact on the electricity market. Peak power demand could be deteriorated without investment in charging infrastructure at working places and throughout cities.

Public or non-residential private parking facilities are characteristic examples of places where large numbers of EVs can concentrate in short periods of time. In order to avoid system disruptions, the aggregated load in the parking facility should be closely coordinated. Techniques that achieve this are widely known as "smart charging".

Smart charging can be either centralized or decentralized. In centralized approaches, EVs transmit a signal with the required State of Charge (SOC) and the desired parking duration, and a control unit allocates charging times and sends the information back to the charger/inverter of each vehicle [3–7]. In decentralized smart charging, the information on spatiotemporal demand is transmitted in the form of incentives that are processed by the in-vehicle controller, which optimizes the charging intervals of the vehicle [8–18].

This process can be facilitated by a Charging Service Provider (CSP), which coordinates charging events with the objective of optimizing one or more from the list below: power losses, transformer overloading, system operation costs, generation costs, vehicle integration, the cost of the power supply, costs for individual drivers, balancing demand and supply, and revenue for the CSP.

For this study, it was assumed that CSPs are contracted as intermediate agents to carry out the task of charging and V2G coordination for parking operators in their control area. As a result, they have a threefold role: (a) to provide EV drivers with their desired SOCs at departure time, (b) to exchange electricity in two directions by respecting the local grid constraints from the Distribution System Operators (DSOs) and (c) to maximize revenue for contracted parking operators. If the parking operators were acting directly as Charging Point Managers (CPMs), there would be no need for the intermediate services of the CSP. These services can be classified as Business-to-Business (B2B) and Business-to-Customer (B2C), and they are schematically presented in Figure 1.

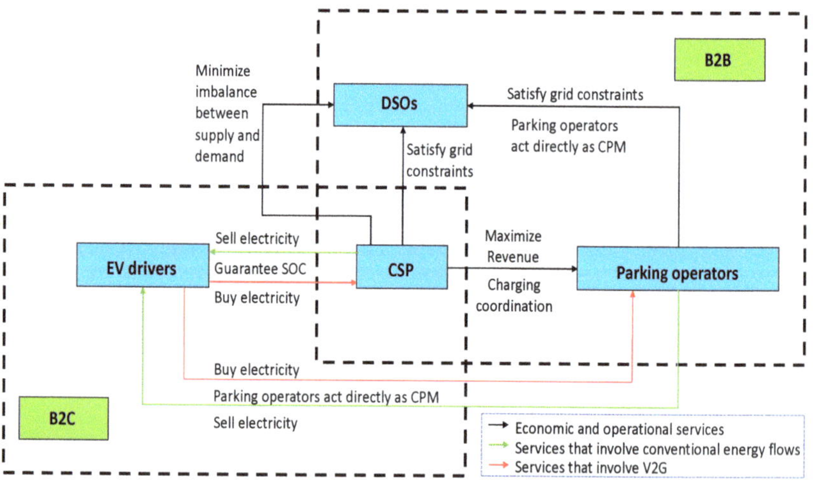

Figure 1. Business-to-Business (B2B) and Business-to-Customer (B2C) services of Charging Service Providers (CSPs) in public and private parking facilities with Vehicle-to-Grid (V2G)-enabled charging infrastructure.

As they grow in size, EV fleets offer greater load flexibility to the CSP. For example, if an individual vehicle needs a lot of energy over a limited period, this energy can be balanced from drivers who are more flexible in their demand and are willing to postpone charging or even provide V2G services. Load flexibility is valuable for charging control, and price incentives can be applied to promote longer charging events with lower power rates and discourage short, power-intense intervals [19].

The work presented in this paper has contributed to the existing literature by developing a Revenue Management (RM) framework for charging coordination. In this framework,

EV drivers reserve, in advance, a parking-and-charging bundle with certain characteristics, such as a charging location, start time and duration, as well as a charging rate. For the users, this approach offers more transparency and control over the charging parameters, and a better understanding of the underlying mechanisms compared to typical scheduling algorithms. For example, if they wish to leave the charging facility earlier, they can estimate, with precision, the final SOC at that time. However, this transition of control from the CSP to the EV driver does not necessarily mean a loss of flexibility. On the contrary, CSPs can vary prices to incentivize low-power bundles; they can better predict EV arrivals, segment their users and optimize their operations in advance.

The modelling framework was evaluated in a microsimulation framework with synthesized activity patterns from a London-based travel diary and choice parameters adopted from a stated preferences experiment [20]. Different EV penetration rates were simulated for a commercial and a shopping area in the city centre.

The rest of the paper is structured as follows: Section 2 discusses the literature review that is relevant to the context and the methodological approach of this study, while the proposed optimization algorithm is presented in Section 3. In Section 4, a brief overview of the data sources that were used for this study is provided along with a demonstration of the simulation framework. Section 5 presents the results and a comparative analysis. Finally, the outcomes of the study are discussed in Section 6.

2. Literature Review

2.1. State of the Art in EV Scheduling and V2G Optimization for Non-Residential Facilities

Several studies have investigated the optimal charging of electric vehicles and the provision of V2G services for peak power and ancillary markets [21–33]. In these studies, the objectives of the control algorithms vary significantly between maximizing economic factors and minimizing the impact on the power network.

The addition of V2G services on top of regular charging complicates the decision process for EV owners. The fast recovery of SOC is desirable because it means that smart charging does not interfere significantly with the daily schedules of the owners. On the other hand, arbitrage techniques based on wholesale electricity prices can generate profits for them at the expense of recharging speed [4].

Rotering and Ilic [21] present a dynamic programming approach to solving this V2G optimization problem, where the objective is to maximize the profit for the EV owners and the decision variable is the daily SOC curve. Zhang et al. [22] examined the potential application of V2G for ancillary services by developing optimal methods for voltage regulation and reactive power control. Shokrzadeh et al. [23] present optimal scheduling strategies for minimizing harmful operating conditions for distribution transformers at a neighbourhood level. DeForest et al. [24] present a centralized method for optimizing EV charging and bid capacity via V2G while maximizing the profit and minimizing the operation costs for the aggregator. Vandael et al. [25] propose a multiagent system for charging coordination with the objective of minimizing imbalance in a smart grid.

Non-residential parking facilities such as parking garages or supermarket parking lots are characteristic examples of places where large numbers of EVs can concentrate in short periods of time. As a result, there is an increasing volume of studies that are shifting their focus away from home recharging.

Yao et al. [26] investigated the optimal integration of EVs in a parking station with the distribution grid according to two different objectives: an economic and a technical one. Shafie-khah et al. [27] present an optimization problem for a parking operator that satisfies demand while curtailing loads for a power utility. They suggest that a constant power rate over the charging session, which is in line with the present paper implementation, can prolong the battery service time. Zhang et al. [28] also modelled the parking operator as a demand response aggregation agent, but they found the optimal level of participation in a set of demand response programs (price-based and incentive-based) instead of focusing on one. Su and Chow [29] developed an intelligent energy-management system for EVs that charge at a municipal parking site, which they solved using probabilistic model-building genetic algorithms.

Finally, the studies that explore the interaction between utilities and parking lots, when V2G is available, are more limited.

Mehta et al. [30] implemented a water-filling algorithm to minimize the variance of the load profile for workplace car parks as a means of reducing peak demand. Babic et al. [31] propose a framework where the parking operator is a broker with two distinct roles: (a) an energy retailer and (b) a player in a target electricity market. The electricity trading functionalities of the operator were evaluated within an agent-based simulation. Moradijoz et al. [32] used genetic algorithms to solve a multi-objective algorithm that optimizes the sites and sizes of parking lots that provide V2G services to the grid.

Probably the research work most closely related to the present paper is that of Hashimoto et al. [33]. In their study, drivers could reserve charging and V2G services through an auction-based system, and the improvements in revenue for the operators were evaluated. It also presents similar methodological assumptions such as discretizing hour intervals for parking reservations and billing by parking duration. In addition, a probability distribution for parking users' willingness to pay was derived from the completion of a questionnaire addressed to them. Finally, the arrival and departure patterns were modelled according to real parking data.

Summarizing the state of the art in EV scheduling methodological approaches, while much effort has been devoted to assessing the economical and network impact of charging and V2G, the current practice largely relies on simplistic representations of charging and parking demand.

2.2. Representation of Charging Demand

In the majority of relevant studies, as was also pointed out by Daina et al. [34], smart charging appraisals adopt predefined charging scenarios and exogenous EV use patterns. The attributes that affect the charging process (the start and end times of charging sessions, initial SOC, subsequent trip duration, parking duration, etc.) are calculated by drawing values from typical probability distributions. Fazelpour et al. [35] used probability distributions of arrival rates and arrival times in a movie theatre parking lot in Tehran, in order to optimize the charging rates of the vehicles. Vandael et al. [25] used fixed charging scenarios, and they assumed that the charging rate is constant during the charging process.

In some rare cases [28], these values have been validated with historical parking information. When real-world charging data are available and are used to deduce charging flexibility [29], these data are not adequate for capturing the elasticity of drivers to charging parameters, because their choices are constrained by the limitations of the provided options.

Recently, there have been some efforts to predict charging behaviour with machine-learning methods. The impact of accurate predictions on charging scheduling has been demonstrated in [36], where the authors suggest a potential 27% decrease in peak load. However, when using historical data of parking durations and energy consumption to predict user behaviour, there is a lack of understanding of individual users' preferences for service attributes, such as the location of the charger and the charging rates.

The prominent novelty of the choice-based pricing optimization that was developed in the present paper is the representation of the charging behaviour in a random utility context and the use of parameters that were empirically estimated from user-tailored choice experiments for charging choices. In a previous study by the authors, the results from these experiments were used under expected and non-expected utility frameworks to understand how people perceive price probabilities and how risk averse they are when they book charging events in advance [37].

Our model bridges the gap in the literature by capturing the heterogeneity in activity-travel and charging preferences and transferring this disaggregate information to a price-based control mechanism.

One of the first studies that modelled the implication of discrete choice models for smart charging services was [34], but in the context of home-based charging activities. In [8,11], the authors developed an activity-based microsimulation where the electric vehicles were controlled with smart charging. Nevertheless, the willingness-to-pay assumptions are not backed by empirical estimation based on revealed or stated preference data. The parameters in the utility function are somewhat arbitrarily tuned using the difference between the forecasted electricity price for next day and the current equivalent price of gasoline. This trade-off makes sense in the decision process for a PHEV driver who can run in both electric and gasoline modes, yet it does not adequately capture the behaviour of a BEV driver.

The integration of charging coordination with the demand response of EV users is achieved through the implementation of revenue management. Before proceeding to the methodological framework in Section 3, there is a brief review of RM and its existing applications in the context of parking and charging.

2.3. *Revenue Management for Electric Vehicle Charging*

Revenue management is a widely adopted method for the allocation and pricing of non-storable services and perishable goods in the service industry. It made its first appearance in the 1970s, when airline companies were deregulated. Data analytics were adopted to differentiate the fares for seats located in the same cabin, and the operators started making dynamic decisions based on predicted demand [38]. Today, we encounter RM techniques in car rental services, hotels, hospitality and most of the industries where there is an inventory with capacity constraints [39].

The first revenue-management approach for a car parking lot was presented in Guadix et al. [40]. In a later study, car parking revenue maximization was achieved by finding the optimal balance between early subscriptions and last-minute drive-in users [41]. In [42], the authors pursued the goal of maximizing revenue for a parking lot in a slightly different manner. A fuzzy-logic-based intelligent parking system with learning capabilities decided, in real time, which reservations to accept and which to reject.

To the best of the authors' knowledge, the first study where revenue management was applied for electric vehicle charging was that of Flath et al. [43]. The optimization objective was to minimize disruption at the distribution network level and to balance energy demand with supply. In this approach, there were two main differentiations from conventional RM methods: (a) instead of discrete inventory units (e.g., hotel rooms, airplane seats, etc.), the energy provided to the EV drivers was continuous, and (b) limited and high-value transactions were replaced by frequent and low-cost ones. In terms of customer segmentation, it is assumed that there are two types of users: drivers who charge their vehicles on a regular schedule (e.g., at home or work) and drivers who have a spontaneous need for topping-up their batteries.

Compared to [43], our methodological approach adds the physical dimension of charging-post availability, moving from a single-resource (energy amount) to a dual-resource allocation. Most importantly, it uses a sophisticated representation of charging demand, and it enables customer segmentation using both quantitative and qualitative choice parameters.

3. Choice-Based Price Optimization

3.1. *Charging Offer Set*

Charging services in the rest of the paper are represented as "bundles" that combine a parking place with the electricity for recharging. As with existing revenue-management applications, this study developed an online reservation system where EV drivers can book their charging bundles up to 1 h before arrival. This system should display all the available options at the time of reservation, including their prices and other service attributes. By packaging out-of-home charging with other parking services, overall prices could become even more appealing than home energy tariffs and attract sufficient charging events to boost investment in public infrastructure.

This is not the first study where users have had the option to choose between different charging offers. In [44], the authors suggest a menu-based pricing system where the users select among contracts of fixed energy quantity and time windows.

The charging bundle offer set for the EV customers was designed by the CSP, and it could either include or not V2G services. The optimal pricing algorithm that is presented here used as input the demand for parking and charging along with the elasticities to the characteristics of each bundle. Given that the size of the price menu escalated exponentially with additional hours of operations, multiple periods of four-hour slots were evaluated.

Moreover, the various bundles were categorized according to their charging rates, which ranged between 3 and 12 kW. It was decided that rapid chargers should not be included in the analysis, due to their relatively low availability at the time that the research was undertaken. The charging preferences of EV customers were estimated under a Latent Class (LC) model.

The optimization problem has two capacity constraints. The first one is the physical constraint of the available charging posts across the parking lots. The second one is the power constraint that is defined by the remaining capacity available to the DSO. While the objective of the algorithm is revenue maximization for the CSP and the contracted parking facilities, the demand-driven management of charging events has the potential for peak shaving and the alleviation of bottlenecks in the local distribution network.

One of the ethical concerns around dynamic pricing is the fairness of the prices. For this reason, the fairest approach is to differentiate prices for EV services based on the impact they have on the grid. Therefore, we need to make sure that we apply accurate bounds for each charging bundle. In order to achieve that, the following equation was formulated:

$$p_j^* = p_{hour} + p_{base} R_j T_j CD_j Ar_j \tag{1}$$

where p_j^* is representative of the specific charging service, and the upper bound is defined at 1.5 p_j^*; p_{hour} is the parking price for an hour, p_{base} is the price for baseload electricity, R_j is the factor that penalizes power-intensive services, T_j is the factor that penalizes peak-load time intervals, CD_j is the charging duration and Ar_j is the factor that penalizes parking lots with high occupancy. All these factors were normalized in a way meaning that the electricity price varied between the base price (10 p/kWh) and a maximum of 55 p/kWh.

In the UK power market, after the actual delivery of electricity, the differences between demand and supply are resolved with an imbalance settlement. This settlement recovers the costs for the system operator by compensating every entity that produced an energy surplus and charging every entity that produced a deficit [45]. Therefore, if the CSP agrees with the DSO for a certain amount of power supply, and the actual demand for EV charging is lower than the expected one, each unit of deficit will have to be reimbursed according to the market index for imbalance costs.

Imbalance prices should be higher for peak-load periods because these are translated to higher costs for the TSO; as a simplification of the complexity of real-market trading values, the imbalance price for each charging bundle was estimated using the T_j factor from Equation (1). Subsequently, these prices were used to calculate the costs from excess power capacity, which were subtracted from the charging service profits to calculate the net revenue for the CSP.

The analysis went a step further by incorporating V2G services in the bundles offered by the CSP. Preferences for V2G services were not estimated because there was a high risk of compromising the estimation validity by augmenting the stated preferences experiments conducted in [20] with V2G scenarios and increasing the complexity for the respondents. Therefore, an assumption was made that the sensitivities to the selling price and discharging amount are identical to those for the buying price and charging amount. The only difference was that the marginal utilities would then have the opposite signs.

While charging and V2G behaviour are assumed to be symmetric, it is likely that this is not the case in reality. For example, the marginal disutility from discharging could be higher for the drivers because of the degradation of the battery or range anxiety. As there are increasing examples of V2G trials around the world, future research could explore these behavioural nuances.

When V2G services are provided, the offer set is extended from 46 to 60 charging bundles. Therefore, EV drivers with low energy requirements (<1 kWh/day) are presented with an extended choice set, where all 14 discharging alternatives deliver a 6 kWh discrete energy quantity to the grid, using different combinations of discharging rates and plug-in durations.

As Moradijoz et al. [32] elaborate, the revenues from V2G power depend on the type of the electricity market that it is sold to. For example, there are markets that pay for energy such as the peak-power market and markets that pay for the available capacity and only require having the vehicle plugged in, such as ancillary services [4]. In the following analysis, only peak-power services were taken into account.

In the next section, we show that by incorporating a latent class model, within the choice-based formulation, it is possible to capture the taste heterogeneity among EV users.

3.2. Utility Specification

The utility of the individual user n selecting the charging bundle j under a discrete choice model specification is the following:

$$U_{jn} = ASC_j + \beta_X X_{jn} + \beta_{XY} X_{jn} Y_n + \beta_{XZ} X_{jn} Z_n \tag{2}$$

where ASC_j is the alternative specific constant of the charging bundle j, β_X is a set of parameters to be estimated, X is a vector of charging-bundle-specific attributes, β_{XY} and β_{XZ} are sets of parameters for interaction terms, Y is a vector of the travel and charging patterns of the individual n, and Z is a vector of individual attributes of the decision maker such as age, marital status, employment type, income, gender and parental status. In particular, the vector X consists of the following characteristic parameters of the charging service:

- The energy required by the user n, E_{jn};
- The parking and charging price, p_j^*;
- The walking time from the charging post to the location where the activity of the individual takes place, WT_{jn};
- The Charging-Induced Schedule Delay Early ($CISDE_{jn}$) and the Charging-Induced Schedule Delay Late ($CISDL_{jn}$).

The last two parameters (i.e., $CISDE_{jn}$ and $CISDL_{jn}$) are based on the theory behind time-of-travel-choice modelling. In particular, the methodology developed in Vickrey's seminal paper [46] suggests that when a commuter choses what time to leave for work, this decision comes after a trade-off between travel time and the measures Schedule Delay Early (SDE) and Schedule Delay Late (SDL). These two measures were defined as follows:

$$SDE = \max\,(PAT - (t_d + TT(t_d)),\ 0) \tag{3}$$

$$SDL = \max\,(t_d + TT(t_d) - PAT_b,\ 0) \tag{4}$$

where PAT is the preferred arrival time, t_d is the time of departure from home and $TT(t_d)$ is the travel time, which is a function of the departure time. We defined $CISDE_{jn}$ and $CISDL_{jn}$ in a similar fashion to the disutility of starting the activity earlier or later, respectively, due to the starting time of the charging event. To better clarify these terms, they are visually demonstrated in Figure 2 for a hypothetical daily scenario.

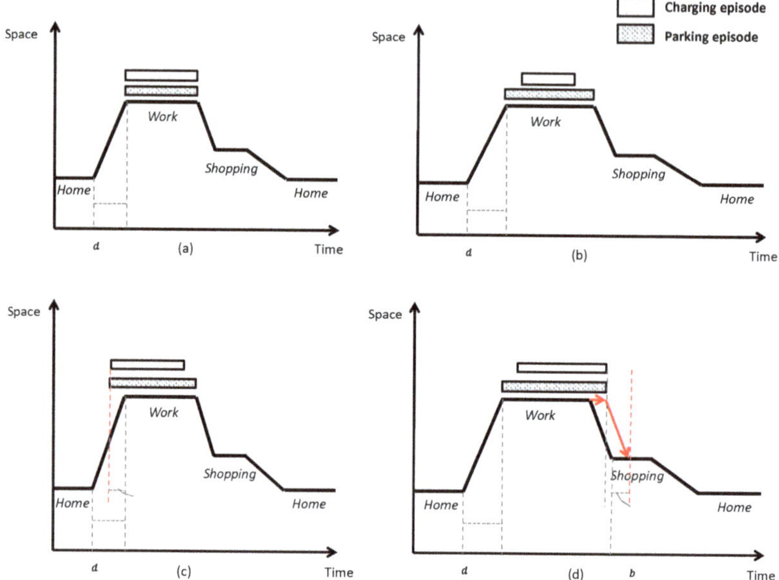

Figure 2. Graphical representation of charging-induced schedule delay for four scenarios: (**a**) there is no schedule disutility, as the charging and the parking episodes are identical; (**b**) there is no schedule disutility, as the charging episode is a subset of the parking episode; (**c**) there is Charging-Induced Schedule Delay Early (CISDE) because both episodes start before the Preferred Arrival Time (PAT); and (**d**) there is Charging-Induced Schedule Delay Late (CISDL) for the next activity because both episodes finish later than the typical departure time. The red dotted lines in (**c**,**d**) represent the actual arrival times that are different from the preferred arrival times.

The majority of the above parameters were estimated in [20]. The parameter for the energy amount β_E was based on a previous estimation of EV drivers' sensitivity to post-charging SOC [47], and the parameter for CISDL was obtained from the CISDE coefficient using the SDE-to-SDL ratio estimated in another study for London commuters [48].

3.3. Latent Class Specification

Latent Class (LC) models are typically applied for segmentation, since they identify classes of users with distinct choice behaviours. As a result, the objective of these models is to achieve intrasegment homogeneity and intersegment heterogeneity based on a set of attributes. Individuals are attributed to each of the classes probabilistically, by estimating class-membership probabilities. A typical LC model formulation is the following:

$$P_n(j|X_{jn}, Y_n, Z_n, \beta) = \sum_{\kappa=1}^{K} P_n(j|X_{jn}, Y_n, Z_n, \beta_\kappa; \kappa) P_n(\kappa|z_n) \quad (5)$$

where K represents the total number of classes, z_n is the set of attributes that makes the behaviour across the segments distinctive and $P_n(\kappa|z_n)$ is the probability that the user n belongs to class κ, conditional on z_n, widely known as the class-membership probability. It is deduced that $\sum_{\kappa=1}^{K} P_n(\kappa|z_n) = 1$. Additionally, $P_n(j|X_{jn}, Y_n, Z_n, \beta_\kappa; \kappa)$ is the class-specific probability calculated in (6).

$$SP_n(j|X_{jn}, Y_n, Z_n, \beta_\kappa; \kappa) = \frac{e^{U_{jn}}}{\sum_{j=1}^{J} e^{U_{jn}} + e^{U_{0n}}} \quad (6)$$

where J is the total set of bundles and U_{0n} is the utility gained from the "no buy" choice. Since "no buy" was not an option in the choice experiments, this utility was approximated by calibrating the alternative specific constant of Equation (2) in a way implying that a small share of the EV drivers did not buy any of the charging bundles.

The typical formulation for a class-specific probability is the multinomial logit (MNL) model; however, other Generalized Extreme Value (GEV) models, such as nested logit or cross-nested logit, can also be adopted to relax some of the hard assumptions of MNL.

The empirical estimation of the latent class model in [20] identified two latent classes: a class where the common characteristic was the high elasticity to price, and a class that was more sensitive to time coefficients, such as walking time and charging duration. The class-membership model is shown in Equation (7).

$$P_n(\kappa|z_n) = \frac{e^{\delta_\kappa + \gamma_\kappa z_n}}{\sum_{l=1}^{K} e^{\delta_l + \gamma_l z_n}} \quad (7)$$

where δ_κ is a constant that is specific to the k class and γ_κ are the parameters to be estimated.

3.4. Price Optimization

In revenue management, it is commonly assumed that the demand is homogeneous. The introduction of discrete choice models as a means to better capture customer behaviour was first attempted with the Choice-Based Deterministic Linear Program (CDLP) [49]. MNL models were replaced by more sophisticated specifications such as nested logit [50] and LC models [51] in subsequent studies.

The objective of the optimization problem developed in this paper is to maximize the expected revenue of the CSP for each four-hour period. The final solution is a menu-based pricing strategy, which should satisfy the constraints for the two-dimensional capacity (charging-post and power availability). The decision variables of this problem, p_j, are the prices of the 46 (or 60 when V2G is available) charging bundles. The latent-class optimization problem is formulated as follows:

$$\max_{p_j} Revenue = D^{EV}[\sum_{j=1}^{J}\{\sum_{\kappa=1}^{K} P_n(j|X_{jn}, Y_n,\ Z_n, \beta_\kappa; \kappa) P_n(\kappa|z_n) p_j\}] - [Y^c - D^{EV} B \sum_{\kappa=1}^{K} P_n(j|X_{jn}, Y_n,\ Z_n, \beta_\kappa; \kappa) P_n(\kappa|z_n) p_j] p^I \quad (8)$$

subject to:

- Capacity constraints:

$$D^{EV} A \sum_{\kappa=1}^{K} P_n(j|X_{jn}, Y_n,\ Z_n, \beta_\kappa; \kappa) P_n(\kappa|z_n) p_j \leq X^c \quad (9)$$

$$D^{EV} B \sum_{\kappa=1}^{K} P_n(j|X_{jn}, Y_n,\ Z_n, \beta_\kappa; \kappa) P_n(\kappa|z_n) p_j \leq Y^c \quad (10)$$

- Price-policy constraints:

$$p_j^- \leq p_j \leq p_j^+ \quad (11)$$

where D^{EV} is the total number of electric vehicles that arrive at the parking facilities for recharging or discharging, X^c is the total number of charging posts, Y^c is the power in kW supplied to the CSP for the contracted facilities, p^I are the imbalance prices, A and B are the incidence matrices for the two capacity dimensions, and finally, p_j^- and p_j^+ are the lower and upper bounds for the decision variables, which are calculated using Equation (1). The methodological approach is visually demonstrated in Figure 3.

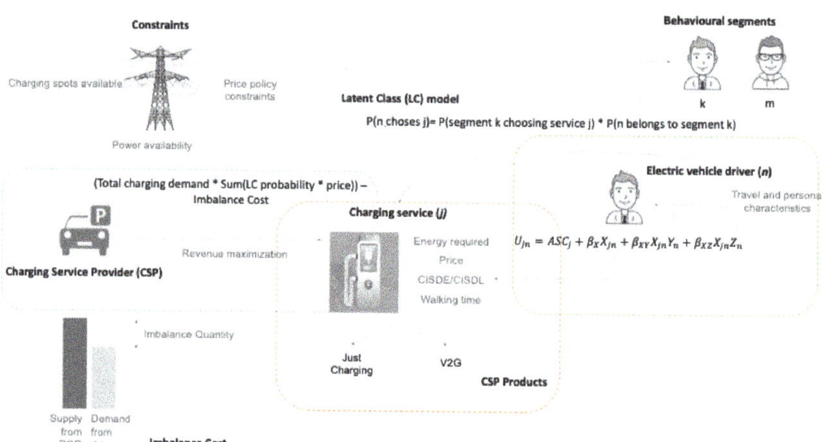

Figure 3. Decomposed elements of the choice-based revenue-management approach. The choice probability for a charging service for an electric vehicle driver affects the revenue outcome for the CSP. Simultaneously, the price, which is the decision variable of the optimization problem, can alter the choice probability, creating a closed-loop formulation with a nonlinear objective function.

Equation (8) can be decomposed in two terms. The first term is the generated profit that is calculated by multiplying the number of EVs by the latent class probability of choosing a charging bundle j and the price of this bundle. The second term is the imbalance cost that is calculated by multiplying the imbalance price for each hour slot with the respective deficit power. The maximum revenue for the CSP is equal to the difference between the generated profit and the imbalance cost for the optimal price menu.

The capacity constraints (9) and (10) correspond to the number of charging posts and the supplied power, respectively. Price-policy constraints (11) vary across charging and V2G services. The minimum price for charging bundles is equal to the maximum price for V2G bundles, and it is assumed to be zero. On the other hand, the maximum prices for charging bundles and minimum prices for V2G bundles reflect their main characteristics; i.e., power-intensive bundles are allowed to have higher prices.

4. Data and Simulation Approach

A simulation approach was employed for the demonstration of the developed pricing algorithm. Two areas with distinct activity patterns and increased travel demand levels were selected for the simulation: a large shopping mall (Westfield Shopping Centre) and a busy commercial area (Canary Wharf). Similar assumptions were made in Battistelli et al. [52], where two garages with EV parking spaces were modelled, serving an office and a residential area. The characteristics of the trips for these areas were extracted from an annual household survey, which combined personal and household information with data from travel diaries: the London Travel Demand Survey (LTDS) [53].

The number of parking spaces used for the simulation corresponded to the existing off-street infrastructure within a 1 km^2 radius of the centroid of each area. It was assumed that these spaces and the hypothetical charging posts were concentrated in two large parking lots for both areas. In terms of data preprocessing, the steps below were followed:

- The units of analysis were tours that started and ended at the homes of the respondents. The tours contained multiple trips.
- Tours that were not identified as car driving, or did not have an intermediate stop within the examined areas, were removed from the analysis.

- If the number of stops in the area of interest was higher than one, the parking event with the highest duration was identified, and that is where the charging event was assumed to take place.

All the electric vehicles in the simulation were assumed to be BEVs, because of range anxiety and their higher likelihood of depending on out-of-home charging infrastructure. At the time of the research, one of the most competitive BEVs in the market was the Nissan Leaf; hence, it was used for the estimation of electricity consumption [54]. For combined city and highway driving, this was set equal to 30 kWh/100 miles, and given a battery capacity of 24 kWh, it corresponds to a driving range of 83 miles. Finally, the energy efficiency of the charging posts was assumed to be 80% and to remain constant at any time step.

The number of charging posts for the simulation was assumed to be equal to the number of parking spaces. While this could be considered a very optimistic scenario for the evolution of electromobility and its infrastructure, there is a rapid deployment of charging posts in large urban centres, which is going to be further facilitated by economies of scale. Furthermore, it should be noted that the areas examined are of high economic interest, with the potential to attract infrastructure investments.

Along with travel demand, these areas have increased electricity demand, especially during hours of peak occupancy. Some examples of appliances that contribute to the aforementioned peaks are personal computers, display lighting and air conditioning units. In order to model the base load of electricity without electric vehicles, first, we identified the baseload profiles for a typical winter weekday for both domestic and nondomestic users. Then, the average population density of residents, the number of employees and the percentages of domestic and nondomestic consumption were used to scale up from a personal profile to the daily distribution. The resulting load curves are presented in Figure 4.

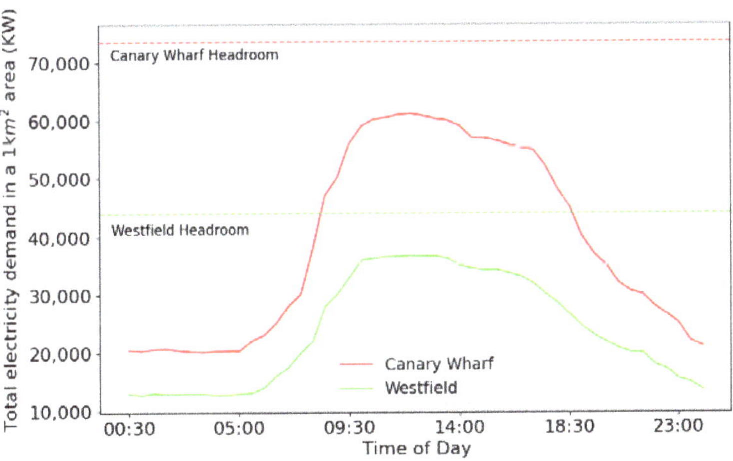

Figure 4. Typical winter weekday baseload curves for the areas of analysis (dashed lines represent installed capacity for the scenario of 20% overload capacity).

The overload capacity can generally fluctuate between local distribution networks. Typically, distribution transformers are replaced when the peak demand grows to be almost equal to the installed capacity. Figure 4 demonstrates the capacities that were selected for the simulation, which in both cases, were 20% higher than the peak value of the base load.

The trip data from LTDS only represent a sample of the population and not the actual demand. Thus, the respondents were used to generate a synthetic population that would allow the exploration of network effects for the two areas. The specifics of the population synthesis algorithm are presented in [20]. The energy requirements for the synthetic EV drivers were calculated based on the reported mileage for all the daily trips. However, since out-of-home charging events were expected to follow a top-up pattern, there was an asymmetric draw from the lower end of the distribution (1–5 kWh).

In order to account for different levels of EV penetration, three scenarios were evaluated: a mid-term scenario (25%), a long-term scenario (50%) and a full-electrification (100%) scenario. As a preliminary step, the incoming demand was satisfied with an uncoordinated charging strategy. This allowed an initial estimation of the spatiotemporal allocation; thus, it could be used for an informed pre-allocation of the supplied power capacity amongst the parking facilities. Some basic assumptions for the uncoordinated scenario were the following: (a) recharging starts as soon as the vehicle is plugged in and (b) the charging event has a constant rate and is evenly distributed over the parking dwell time.

Combining the data sources and the synthetic information that has been described so far in this section, we present all the variables that were used for simulation and optimization in this study, in Table 1. Scaling up the trip sample dataset using aggregate statistics for the areas of interest, we ended up with 10,852 trips for Canary Wharf and 14,360 trips for Westfield. Then, depending on the scenario for EV penetration levels, the total numbers of trips in the simulation are presented at the end of the table.

Table 1. Description of the input data used in the simulation.

Description	Value Range	Description	Value Range
Parking and charging characteristics		**Demographics**	
Parking start time	9:00–20:00	Gender	Male or female
Parking end time	10:00–21:00	Age	<=40 vs. >40
Parking location	1 or 2	Marital status	Married vs. all other categories
Constant charging rate (in kW)	0.5–12	Employment status	Employed vs. unemployed
Walking distance from activity to parking locations (in m)	10–1700	Income (in GBP)	<GBP 10,000 to >100,000
Energy requirements (in kWh)	0–24	Children in the household	Yes or no
Charging bundle characteristics		**Context variables**	
Discrete charging rate (in kW)	[−6,−3,3,6,8,12]	Day of week	Monday–Sunday
CISDE (in minutes)	0–180	Initial charging prices (GBP/kWh)	0.10–0.55
CISDL (in minutes)	0–180	Energy consumption (kWh/miles)	30/100
Charging duration (in minutes)	0–240	Battery range (in miles)	83
Number of products (no V2G/V2G)	46/60	Charging efficiency	80%
		Number of charging posts	625
		Work-based tour	Yes or no
		Distribution network headroom	20%
Size of synthetic daily trip dataset			
	Canary Wharf		Westfield
25% EVs scenario	2713		3590
50% EVs scenario	5426		7180
100% EVs scenario	10,852		14,360

The first step of the methodological approach was to examine an uncontrolled charging scenario. The parking and charging characteristics in the top left of Table 1 were deduced from the trip dataset following a set of assumptions and rules. For example, it was assumed that the driver will park at the facility that is closer to the final destination and that the energy requirements will depend on the subsequent trips of the day and the remaining SOC. Subsequently, the initial SOC depends on the distance driven so far and the energy consumption. Additional context variables that were necessary for the simulation are depicted in the middle-right part of the table.

The "dumb charging" strategy is useful for understanding the spatiotemporal allocation of demand and applying the simple area-based and time-based fixed prices that are explained in Equation (1). It also enables the higher allocation of power capacity to the busiest parking facility as a strategic decision. In the next step of our methodology, the charging events were driven by actual choices of the users, which in turn, were probabilistic outcomes of a decision process. Individuals try to maximize their utility (Equation (2)), which is a linear combination of factors from all the sections of Table 1. Using the choice-based formulation described in Equation (8), the prices of the charging bundles were optimized with respect to CSP revenue. Then, they were used to rerun the simulation and evaluate other key metrics such as the load factors and demand–supply imbalance.

EV scheduling problems are typically characterized by large numbers of variables and constraints that are not continuously differentiable and increase the related models' execution times for finding an optimal solution. Metaheuristic algorithms (MAs) are very popular in the relevant literature as a means for solving these NP-hard (nondeterministic polynomial-time hard) problems [55,56]. The main disadvantage of MAs is that they are not guaranteed to reach the global optimum, due to the stochasticity in the process. However, V2G scheduling problems are typically highly dimensional, nonconvex and nonlinear optimization problems, and MAs are arguably one of the best options for both binary and real-valued problems.

The formulation in this paper is, in fact, a constrained nonlinear problem, and the two main categories of MAs that are applied to solve such are Genetic Algorithms (GAs) and Particle Swarm Optimization (PSO).

The reader is referred to [55,56] for a detailed review of GAs, PSO and other metaheuristics applied in Unit Commitment (UC) problems and EV charging/discharging coordination. The studies [57,58] are typical applications of the two methods. In [57], the authors used a GA in a game theoretic analysis of EV charging coordination. On the other hand, Hutson et al. [58] applied binary PSO to find the optimal buying and selling times for a fleet of vehicles in a parking lot.

The genetic algorithm creates a population of candidate price vectors for the charging bundles, and the best candidate approaches the optimal price vector. Each population is generated after applying certain stochastic operators to the previous population. In particular, the revenue for the operator, which is the fitness score in this application, is calculated for every price vector in the population using Equation (8). If the score of the vector is higher than the average fitness score, it is selected to be transitioned to the next population. This process is typically known as selection. If the score is lower than the average fitness score, random changes are applied to the parent price vectors in order to generate children for the new population. This process of adding diversity in the creation of new offspring is typically known as mutation. If some of the prices in a price vector lead to solutions that violate capacity constraints (Equations (9) and (10)) or price-policy constraints (Equation (11)), a penalty score is incremented by the maximum price. In this way, infeasible price candidates are less likely to be selected for the next population.

The algorithm is presented in detail in Figure 5. First, we initialize the number of iterations (generations) as $G = 100$. We randomly generate an initial population P^0 of size $P = 100$. Each individual in the population is represented by a D-dimensional vector, where D is the number of charging (or charging and discharging) bundles:

$$p_i = (p_{i1}, p_{i2}, \ldots, p_{iD}) \tag{12}$$

and the genetic encoding is a direct value encoding the price. This value can be between 0 and the upper bound for the charging bundles or between the lower bound and 0 for the discharging bundles. The fitness function (the choice-based revenue function in Equation (8)) is evaluated for all the individuals in P^0. Then, as long as the stopping criterion is not satisfied, we iteratively create subsequent populations. We calculate the score (revenue) for each individual and the average score of the population. We select the individuals with scores higher than the average population score to move to the next generation, and we remove the remaining ones. We also mutate the genes (prices) of the successful individuals by adding random noise, in order to create new price vectors and always have a population of size 100. At each generation *t*, we evaluate the fitness function P^t, and we follow the same steps. Finally, when the stopping criterion is satisfied, the iteration is terminated, and the individual that generates the maximum revenue is selected.

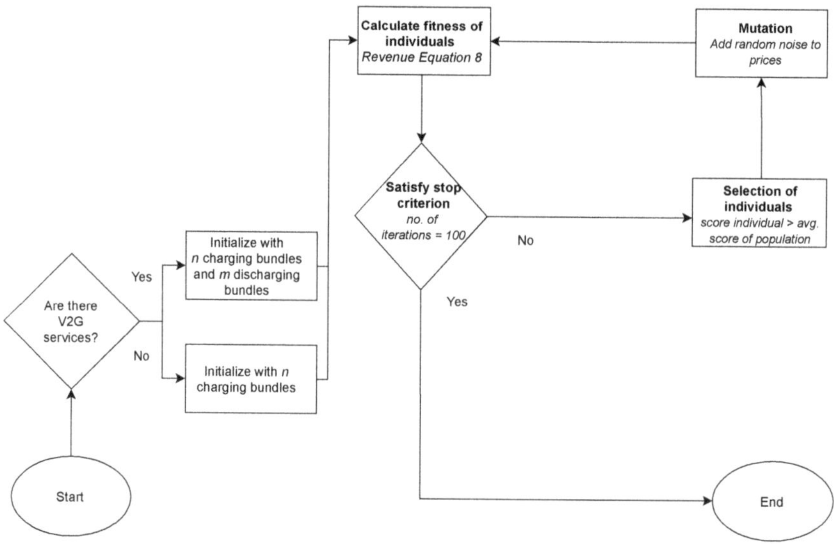

Figure 5. Genetic algorithm solution for price optimization.

The second nature-inspired algorithm that was adopted in this study for optimization is the PSO. Like the genetic algorithm, each individual in a population (here referred to as swarm because it is based on the information exchange of birds in a swarm) is updated in an iterative manner. The exploration of the problem search space is guaranteed by introducing, again, a certain stochasticity in the transitions. Upon initialization, each individual particle in the swarm creates a randomized position solution, i.e., vector of prices, and a randomized velocity within a uniform range of values. Then, the initial positions are assigned to the particles' best-known positions. For each iteration, the algorithm updates the velocity of the price vectors using their best individual positions and the best position of the swarm. A cognitive constant c_1 limits the influence that the particle's best-known positions have on their new velocity, while a social constant c_2 limits the influence the best price vector of the swarm has on the other vectors.

The details of the PSO solution are presented in Figure 6. The algorithm starts by initializing a set of parameters including the acceleration constants c_1 and c_2. The size of the swarm is assumed to be $P = 100$. The position of each solution (particle) i of the swarm is represented by a D-dimensional vector, where D is the number of charging (or charging and discharging) bundles, as depicted earlier in Equation (12). Simultaneously, each particle is randomly assigned an initial velocity, which is given by:

$$v_i = (v_{i1}, v_{i2}, \ldots, v_{iD}) \tag{13}$$

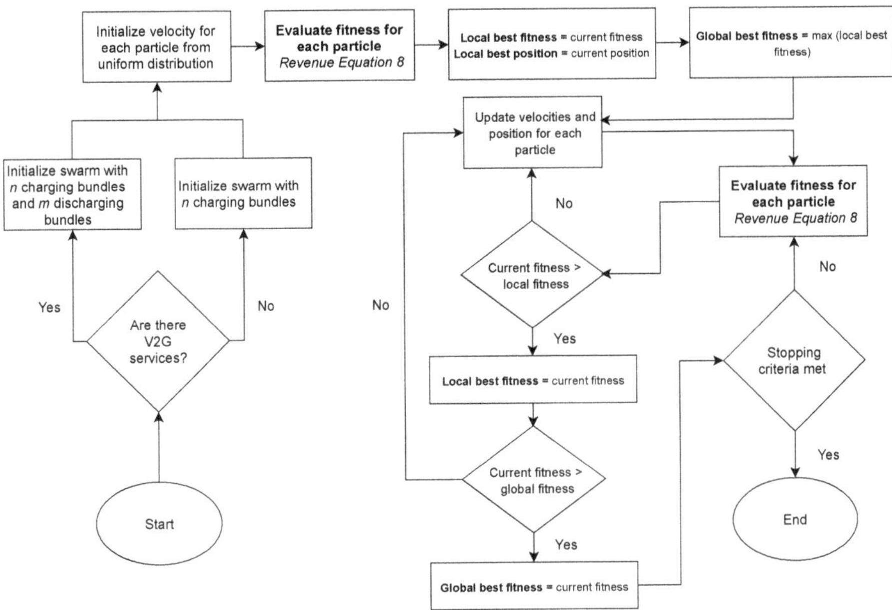

Figure 6. Particle Swarm Optimization (PSO) solution for price optimization.

The fitness value (Equation (8)) is evaluated for each particle's position $f(x_i)$. The current position is assigned as the best local position (x^*_i), while the position with the highest fitness value is assigned as the best global position (g_{best}). Then, as long as the termination criterion is not satisfied, the positions and velocities of the swarm particles are updated based on (a) their own best local positions and (b) the global best positions in their neighbourhood:

$$p_i^{(t+1)} = p_i^{(t)} + v_i^{(t+1)} \quad (14)$$

$$v_i^{(t+1)} = v_i^{(t)} + c_1 r_{i1}\left(x_i^{*(t)} - x_i^{(t)}\right) + c_2 r_{i2}\left(gbest - x_i^{(t)'}\right) \quad (15)$$

where t is the iteration count, while r_1 and r_2 are random vectors that take values between 0 and 1. The fitness value is evaluated for the new position, and the local and global best solutions are updated. At the end, when the termination criterion is satisfied ($t = 100$), the best particle is selected as the optimal price vector.

Both the GA and PSO algorithms were developed in Python, which was also used to build the simulation framework.

The simulation framework, which is demonstrated in Figure 7, considered four control scenarios: (a) fixed pricing based on the time of day (FP), (b) fixed pricing based on the time of day and typical spatial demand (FP2), (c) optimal pricing with the GA, and (d) optimal pricing with PSO. All these control scenarios were compared with an uncontrolled charging scenario where it was assumed that charging demand was equally allocated for the parking duration

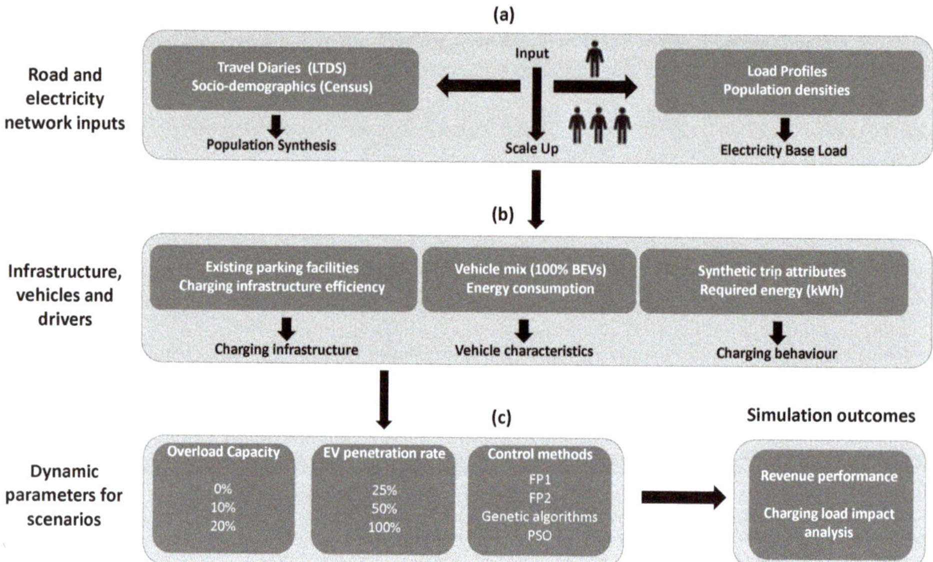

Figure 7. Simulation framework for charging and V2G coordination. The three levels capture (**a**) the synthetic approach to building a representative demand for travel and energy, (**b**) the technology and behavioural assumptions and (**c**) the dynamic parameters that were evaluated under different scenarios.

All the simulation runs account for a 12 h operating window between 9:00 and 21:00 with overlapping subsequent 4 h scheduling windows. By running the simulation for all the possible combinations, a total of 48 cases were modelled. Breaking down the problem into 4 h subproblems led to near-optimal solutions but, simultaneously, did not become computationally expensive, which would be prohibitive for the numerous scenarios that were analysed.

5. Results

The main metrics that were evaluated after each simulation run are the revenue for the charging service provider, and the load factors for both the physical dimension of charging posts and the power supplied by the DSO.

Figures 8 and 9 demonstrate the results for the simulation across two dimensions: the percentage of EVs and the overload capacity. Figure 8 corresponds to the heavy business area, while Figure 9 corresponds to the commercial area. For each case study, we examined the load curve against a transformer capacity that was designed based on the maximum value of the baseline curve (first column) and on an overload capacity of 20% (second column). Finally, the three rows are associated with an increasing percentage of EVs from top to bottom.

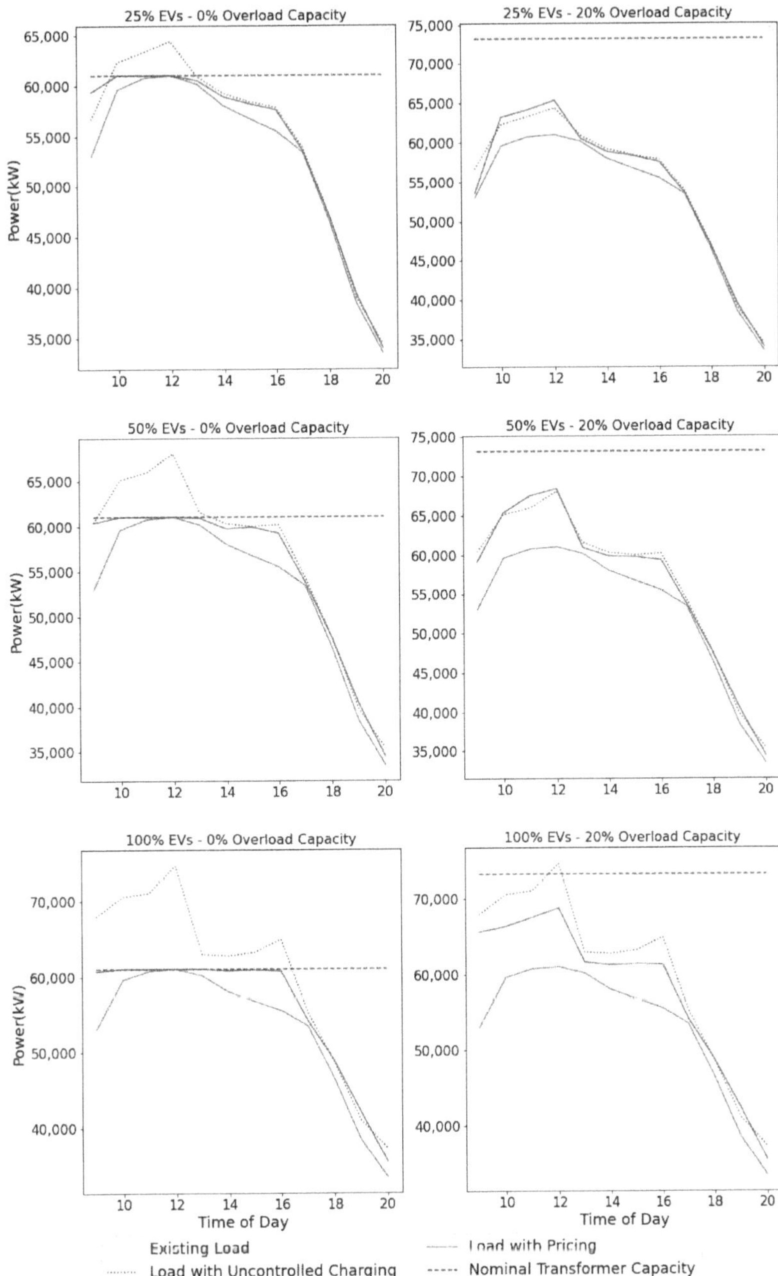

Figure 8. Charging allocation analysis for Electric Vehicle (EV) market penetration and overload capacity (Canary Wharf area). The red dotted line represents the nominal capacity of the distribution transformer; the grey line, the baseline load demand; the blue dotted line, the uncoordinated charging scenario; and the green line, the charging allocation with Fixed Pricing based on time of day (FP).

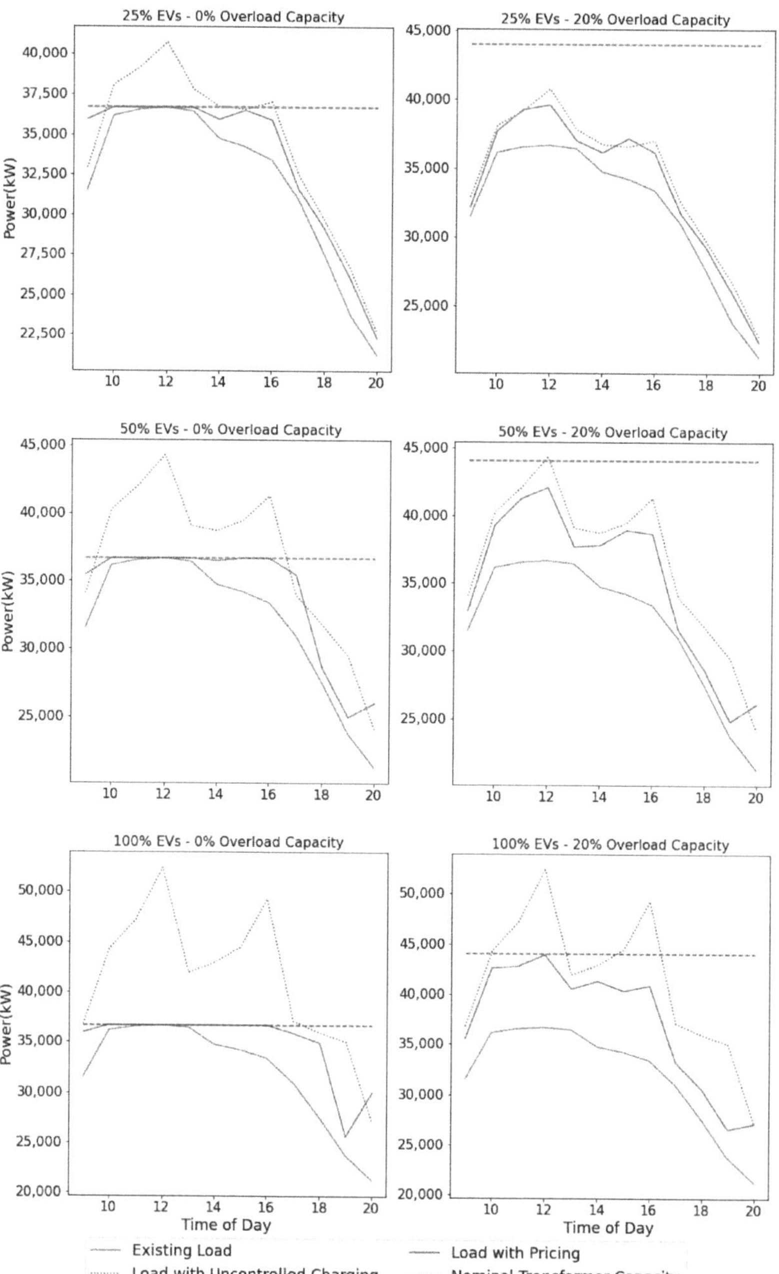

Figure 9. Charging allocation analysis for EV market penetration and overload capacity (Westfield Shopping Centre area). The red dotted line represents the nominal capacity of the distribution transformer; the grey line, the baseline load demand; the blue dotted line, the uncoordinated charging scenario; and the green line; the charging allocation with Fixed Pricing based on time of day (FP).

Since the peak demand for the study areas is already high, a 20% overload capacity is a significant increase in the available power, and it is sufficient, even for the full-electrification scenario. The baseline load distribution is quite similar for the two areas, with the busiest time being around midday. For the uncoordinated-charging scenario, there are several periods where demand exceeds the available capacity. The most extreme case is the scenario for Westfield, with 100% of EVs and 0% overload capacity.

The FP control algorithm takes into account historical driving and activity data to penalize the hour slots with the highest charging demand. The price for each charging bundle is calculated based on Equation (1) without the area factor. Each charging bundle is allocated only if it is not constrained by power availability and the number of free charging posts. This pricing incentive initiates a behavioural shift, with some EV drivers charging later in the day, while some drivers with low SOC needs are discouraged from charging at all.

Figures 8 and 9 do not include V2G services for parking customers. The impact of V2G services can be observed in Figure 10. From this point on, only the Canary Wharf results are presented because the relative effects of the parameters on the results were similar for the Westfield area. The difference is that now, the overload capacity was kept fixed at 0%, and it is replaced in the graphs by the V2G availability parameter.

It is interesting to observe that the final load curve is below the baseline curve for the majority of the day. This can be explained by the fact that several drivers prefer to sell electricity back to the grid instead of charging their cars, even taking into account the disutility from the reduced SOC for the rest of their daily trips. This is extremely useful for periods of peak demand because it allows other drivers with higher charging needs to refill their batteries. Nevertheless, it incurs additional imbalance costs to the CSP during nonpeak periods, when the V2G services are not required.

The net revenue for the CSP is optimized with the choice-based pricing algorithm that was presented in the previous sections. Figure 11 shows the load curves under the four different control approaches, with and without V2G availability. FP2 is similar to FP, but now, the area factor is included in the calculation of Equation (1). In this way, busy parking is penalized, and drivers are incentivized to plug in their cars in a more distant location.

In terms of overall load scheduling, this method has very similar results to the previous one. On the other hand, the two solution algorithms for our optimization problem have a more profound effect. When V2G is not available, a higher portion of charging events is shifted from peak to nonpeak hour slots by setting prices that are closer to the drivers' willingness to pay. To better understand what happens when V2G services are provided, one can have a look at Table 2.

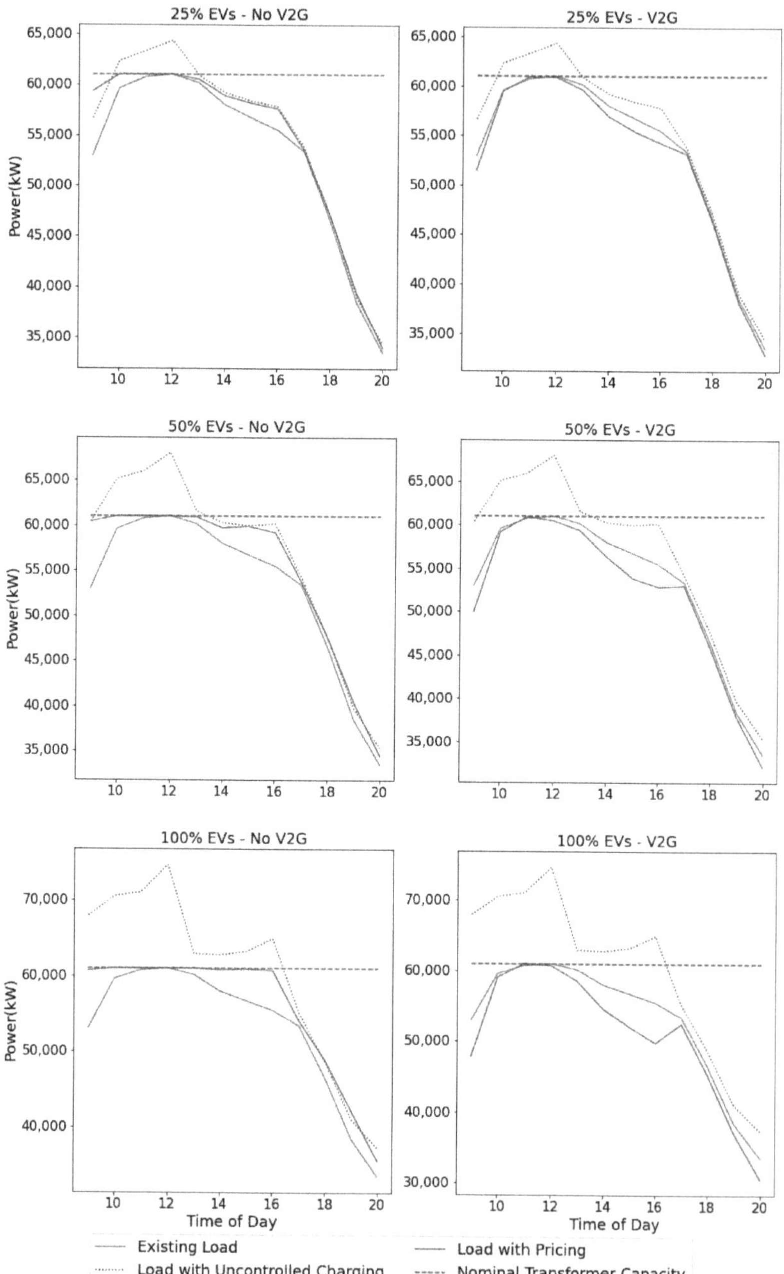

Figure 10. Charging allocation analysis for EV market penetration and V2G availability (Canary Wharf area and 0% overload capacity). The red dotted line represents the nominal capacity of the distribution transformer; the grey line, the baseline load demand; the blue dotted line, the uncoordinated charging scenario; and the green line, the charging allocation with Fixed Pricing based on time of day (FP).

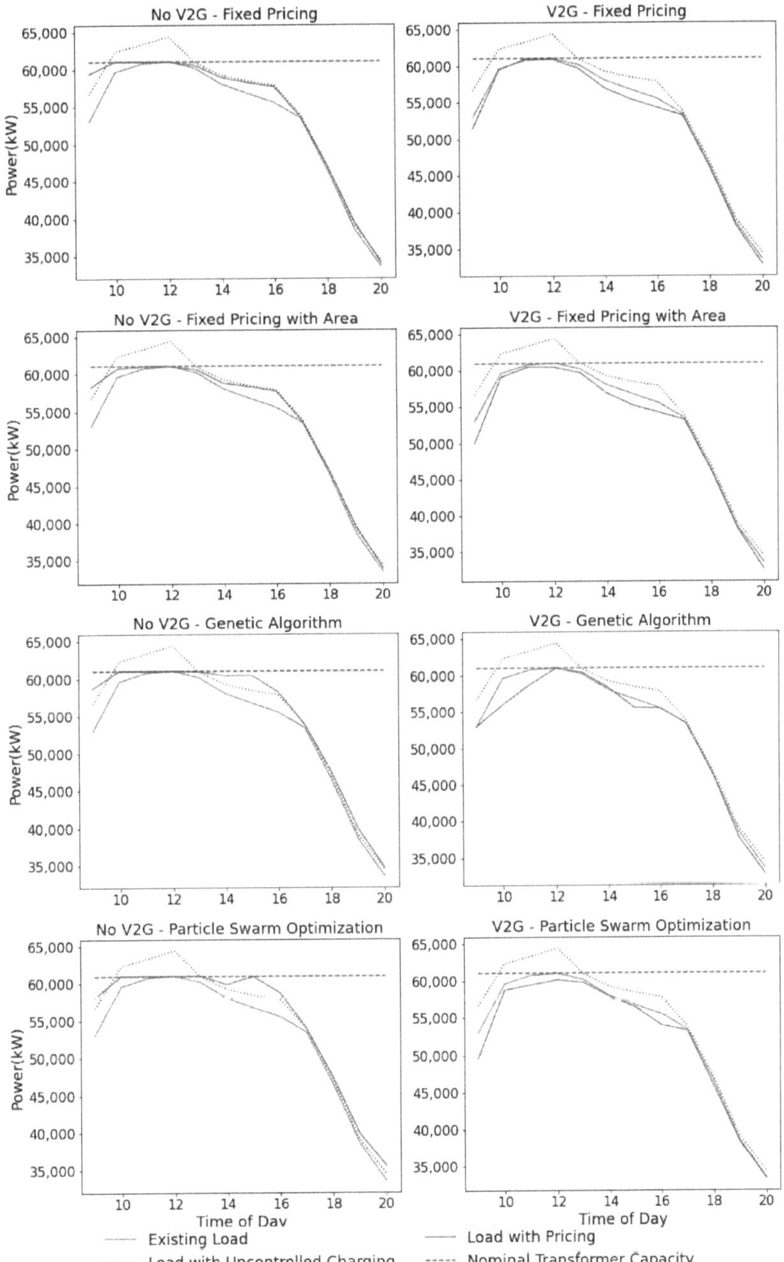

Figure 11. Charging allocation analysis for V2G availability and control algorithm (Canary Wharf area and 0% overload capacity). The red dotted line represents the nominal capacity of the distribution transformer; the grey line, the baseline load demand; the blue dotted line, the uncoordinated charging scenario; and the green line, the charging allocation with the respective algorithm.

Table 2. Revenue performance and parking load factors for various parameter combinations.

V2G Availability	EV Market Penetration	Control Scenario	Overload Capacity	Revenue (GBP)	Parking Load Factor	Parking Load Factor (P1)	Parking Load Factor (P2)
No V2G	100%	FP	0.0	5768	40%	43%	38%
			0.2	7980	69%	69%	68%
	50%	FP	0.0	3775	27%	27%	27
			0.2	5268	50%	51%	50%
	25%	FP	0.0	1959	15%	16%	14%
			0.2	2818	31%	31%	31%
		FP2	0.0	1835	15%	8%	22%
		GA	0.0	6056	15%	16%	14%
		PSO	0.0	8051	15%	16%	13%
V2G	100%	FP	0.0	−5278	72%	73%	71%
			0.2	−5026	72%	72%	72%
	50%	FP	0.0	−2691	48%	51%	45%
			0.2	−2657	50%	51%	48%
	25%	FP	0.0	−1358	25%	27%	23%
			0.2	−1230	27%	28%	26%
		FP2	0.0	−2005	25%	31%	18%
		GA	0.0	−6.86	18%	9%	26%
		PSO	0.0	−1040	20%	20%	20%

When drivers are allowed to sell electricity to the grid using V2G technology, all the methods result in losses for the CSP because the expenses from the "selling" bundles exceed the income from the "buying" bundles. The choice-based optimization reduces the losses, especially when the GA solution is applied. The reduced parking load factors indicate that upon decreasing the sale prices to avoid excessive V2G and reduce the imbalance costs, the V2G activities are spread throughout the day.

The trade-offs between charging and discharging in the choice model heavily rely on the customers' sensitivity to price. As was highlighted earlier, the class-specific parameters for the two latent classes were estimated by the authors in previous work. Therefore, the estimates were bound to the tariffs used in the choice experiments and do not necessarily reflect future fluctuations in electricity prices. The effect of this behavioural uncertainty on the outcome of the optimization problem was addressed by performing a sensitivity analysis regarding the price coefficients. The outcomes are demonstrated in Table 3 for the GA solution. The relative differences were similar for the PSO solution.

Table 3. Analysis of revenue and parking load factor sensitivity to price parameters.

Price Sensitivity	Revenue (GBP)	Parking Load Factor	Parking Load Factor (P1)	Parking Load Factor (P2)
Decreased	850.21	23%	24%	13%
Medium	−6.41	20%	11%	28%
Increased	−791.26	18%	17%	30%

The original specification was defined as "Medium Price Sensitivity", and the price parameters were halved and doubled, respectively, for the "Decreased" and the "Increased" scenarios. When drivers are less sensitive to charging prices, their utility is overruled by their energy needs and their willingness to walk, so there is an increase in drivers that charge in parking lot 1, and the overall revenue becomes positive for the CSP. On the contrary, when drivers are more sensitive to charging prices, the number of V2G events increases in both parking facilities. Consequently, the imbalance costs are increased by the excess electricity, and the optimal solution leads to an overall loss.

The following results indicate the significance of the demand parameters in the developed framework. If similar datasets become available in the future, a cross-validation could be useful since some elements of the choice experiments that were used are not established at the moment (e.g., workplace charging services), and some properties were approximated (e.g., the sensitivity to energy quantity). At the same time, the estimated parameters should be treated with caution when applied to other geographic locations, since it is likely that they are correlated with some idiosyncratic preferences of British drivers.

The next step in the analysis was to understand how the optimal prices were related to the attributes of the charging bundles. Figure 12 demonstrates a classification of the prices that were generated with the GA solution by power and charging duration. It is observed that short-duration, high-power bundles tend to be more dispersed around 0 compared to long-duration, low-power bundles. This is essentially an indirect reward to users that strain the power network less by spreading their charging demand over time.

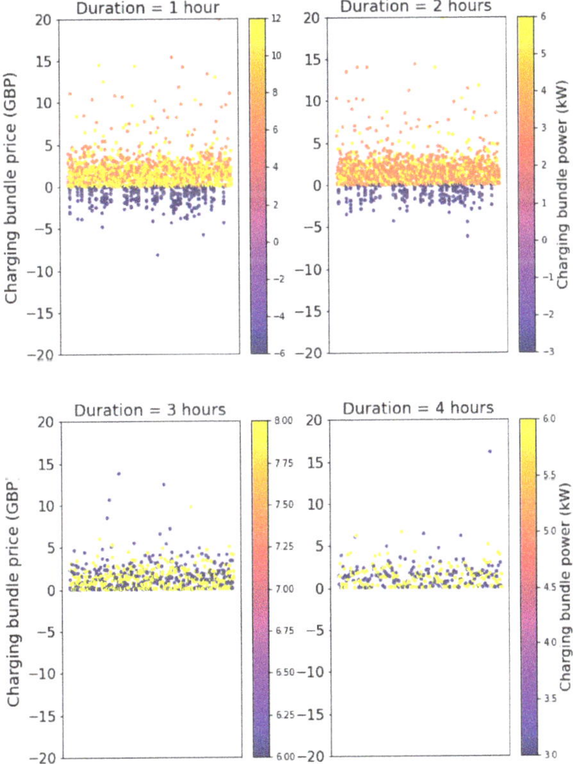

Figure 12. Classification of GA-optimized prices for charging and V2G bundles by power and charging duration.

To conclude this section, we performed a comparative analysis of the two algorithms' performance in terms of computational time. The run time of the optimization was evaluated for three different parameters: (a) the number of iterations, (b) the number of populations (GA) or the number of particles (PSO) and (c) the effect of scaling up the number of EVs. All the parameters were found to have linear effects on computational time as is shown in Table 4.

Table 4. Comparative analysis of computational times for solution algorithms.

Genetic Algorithms				Particle Swarm Optimization			
No. of Iter.	No. of Populations	Scale (%)	Time (s)	No. of Iter.	No. of Populations	Scale (%)	Time (s)
10	10	25	20.1	10	10	25	17.1
10	10	50	40.6	10	10	50	35.0
10	10	100	83.6	10	10	100	69.2
100	10	25	186.7	100	10	25	184.5
10	100	25	192.7	10	100	25	194.3

The metaheuristics' overall performance can be summarized as follows:

- Their computational times are very similar;
- The GA algorithm provides a better solution without V2G;
- The PSO algorithm provides a better solution with V2G.

6. Conclusions

This paper shows how a choice-based revenue-management problem can be integrated with the parking and charging choices of electric vehicle drivers. In particular, an integrated latent class and nonlinear framework was developed to optimize prices for charging services provided by a charging service provider.

There are two innovative elements of the developed methodological framework in comparison to existing research.

First, the endogenous relationship between the sensitivity of EV drivers to charging characteristics and the charging coordination methods applied by the operator was captured with a sophisticated disaggregate model of demand. Most importantly, the parameters of this model were empirically estimated from user-tailored choice experiments for charging choices.

The second contribution of this paper is the development of a framework that can have direct implementation in the charging service industry. Revenue management allows the closed-loop integration of supply and demand and has proven to be financially beneficial in several service industries. The excessive number of papers that have been written in the last decade aiming to accommodate the increasing charging needs of EV drivers cannot always be aligned with practice-ready business models.

A microsimulation framework was used to implement the pricing model for a synthesized network of two distinctive regions. Baseload electricity curves were modelled by scaling up typical load profiles, and survey data for travel behaviour were scaled up by using synthetic population methods. The charging behaviour of the simulated drivers was modelled using an advanced discrete choice model, and different scenarios were established for EV penetration rates, the overload capacity and the V2G availability.

One limitation of this online reservation system is that it cannot capture last-minute stochastic arrivals. As was explained in the introduction, this approach reduces the uncertainty both for the users who need to be reassured of a certain level of SOC for their subsequent trip and for the parking operator who wants to achieve a smooth allocation of charging demand. However, it results in a conservative lower-bound estimation of the optimal revenue. Future research could combine a revenue-management approach, which by definition, has to be resolved in a preservice booking period, with a more dynamic application that considers last-minute arrivals.

The results suggest that the revenue-management framework simultaneously maximizes revenue and assists in the prevention of local transformer overloading. Charging peaks are alleviated especially when V2G services are adopted to provide energy back to the grid. In addition, it accommodates drivers during the peak hours, while before, they were unable to charge their vehicles because of network constraints.

The outcomes of this research could potentially be interesting for retail operators that host charging infrastructure and want to understand revenue opportunities from merging charging and retailing services. The charging bundles described could be extended to incorporate a point system or the exchange of electricity with retail products. At the same time, the power constraints could reflect the energy needs of a retail building, such as, for example, lighting, heating and cooling for a supermarket in high-demand hours. It is significant to highlight here the rapid increase in studies that are exploring the integration of V2G with energy-management systems in buildings [59–61].

Author Contributions: Conceptualization, C.L., A.S. and J.W.P.; methodology, C.L., A.S. and J.W.P.; software, C.L.; validation, C.L.; formal analysis, C.L.; investigation, C.L., A.S., and J.W.P.; resources, A.S. and J.W.P.; data curation, C.L.; writing—original draft preparation, C.L.; writing—review and editing, C.L. and A.S.; visualization, C.L.; supervision, A.S. and J.W.P.; project administration, A.S. and J.W.P.; funding acquisition, A.S. and J.W.P. All authors have read and agreed to the published version of the manuscript.

Funding: This work was partially supported by the Grantham Institute for Climate Change and Climate KIC.

Institutional Review Board Statement: Not applicable.

Informed Consent Statement: Not applicable.

Data Availability Statement: The data presented in this study are openly available in Dryad at https://doi.org/10.5061/dryad.c59zw3r68.

Acknowledgments: The authors would like to thank Transport for London for providing the LTDS dataset that was indispensable for the analysis. We would like to dedicate this paper to the recently deceased J.W.P. for his invaluable inspiration and support towards this research.

Conflicts of Interest: The authors declare no conflict of interest. The funders had no role in the design of the study; in the collection, analyses or interpretation of data; in the writing of the manuscript; or in the decision to publish the results.

References

1. Electric Drive Transportation Association. Electric Drive Sales Dashboard. Available online: http://electricdrive.org/index.php?ht\protect$\relax\protect{\begingroup1\endgroup\@@over4}$d/sp/i/20952/pid/20952 (accessed on 28 December 2019).
2. Ulrich, L. 2017 top ten tech cars. *IEEE Spectr.* **2017**, *54*, 26–35. [CrossRef]
3. Kempton, W.; Tomic, J. Vehicle-to-grid implementation: From stabilizing the grid to supporting large-scale renewable energy. *J. Power Sources* **2005**, *144*, 280–294. [CrossRef]
4. Kempton, W.; Tomić, J. Vehicle-to-grid power fundamentals: Calculating capacity and net revenue. *J. Power Sources* **2005**, *144*, 268–279. [CrossRef]
5. Acha, S.; Green, T.C.; Shah, N. Effects of Optimised Plug-In Hybrid Vehicle Charging Strategies on Electric Distribution Network Losses. In Proceedings of the IEEE PES T&D 2010, Sao Paulo, Brazil, 8–10 November 2010; pp. 1–6.
6. Clement-Nyns, K.; Haesen, E.; Driesen, J. Coordinated Charging of Multiple Plugin Hybrid Electric Vehicles in Residential Distribution Grids. In Proceedings of the 2009 IEEE/PES Power Systems Conference and Exposition, Seattle, WA, USA, 15–18 March 2009; pp. 1–7.
7. Zheng, Y.; Shang, Y.; Shao, Z.; Jian, L. A novel real-time scheduling strategy with near-linear complexity for integrating large-scale electric vehicles into smart grid. *Appl. Energy* **2018**, *217*, 1–13. [CrossRef]
8. Galus, M.D.; Waraich, R.A.; Noembrini, F.; Steurs, K.; Georges, G.; Boulouchos, K.; Axhausen, K.W.; Andersson, G. Integrating power systems, transport systems and vehicle technology for electric mobility impact assessment and efficient control. *IEEE Trans. Smart Grid* **2012**, *3*, 934–949. [CrossRef]
9. Shafiekhah, M.; Heydarian-Forushani, E.; Golshan, M.; Siano, P.; Moghaddam, M.P.; Sheikh-El-Eslami, M.; Catalão, J. Optimal trading of plug-in electric vehicle aggregation agents in a market environment for sustainability. *Appl. Energy* **2016**, *162*, 601–612. [CrossRef]
10. Lopes, J.A.P.; Soares, F.J.; Almeida, P.M.R. Identifying Management Procedures to Deal with Connection of Electric Vehicles in the Grid. In Proceedings of the 2009 IEEE Bucharest PowerTech, Bucharest, Romania, 28 June–2 July 2009; pp. 1–8.
11. Waraich, R.A.; Galus, M.D.; Dobler, C.; Balmer, M.; Andersson, G.; Axhausen, K.W. Plug-in hybrid electric vehicles and smart grids: Investigations based on a microsimulation. *Transp. Res. Part C Emerg. Technol.* **2013**, *28*, 74–86. [CrossRef]

12. Gan, L.; Topcu, U.; Low, S.H. Optimal decentralized protocol for electric vehicle charging. *IEEE Trans. Power Syst.* **2013**, *28*, 940–951. [CrossRef]
13. Ma, Z.; Callaway, D.; Hiskens, I. Decentralised Charging Control for Large Populations of Plug-In Electric Vehicles. In Proceedings of the CDC 2010: 49th IEEE Conference on Decision and Control, Atlanta, GA, USA, 15–17 December 2010; pp. 206–212.
14. Karfopoulos, E.L.; Hatziargyriou, N.D. A multi-agent system for controlled charging of a large population of electric vehicles. *IEEE Trans. Power Syst.* **2013**, *28*, 1196–1204. [CrossRef]
15. Wen, C.-K.; Chen, J.-C.; Teng, J.-H.; Ting, P. Decentralized plug-in electric vehicle charging selection algorithm in power systems. *IEEE Trans. Smart Grid* **2012**, *3*, 1779–1789. [CrossRef]
16. Papadaskalopoulos, D.; Strbac, G. Participation of Electric Vehicles in Electricity Markets through a Decentralised Mechanism. In Proceedings of the 2011 2nd IEEE PES International Conference and Exhibition on Innovative Smart Grid Technologies (ISGT Europe 2011), Manchester, UK, 5–7 December 2011; pp. 1–8.
17. Papadaskalopoulos, D.; Strbac, G. Decentralised Participation of Electric Vehicles in Network-Constrained Market Operation. In Proceedings of the 2012 3rd IEEE PES Innovative Smart Grid Technologies Europe (ISGT Europe 2012), Berlin, Germany, 14–17 October 2012; pp. 1–8.
18. Cao, Y.; Tang, S.; Li, C.; Zhang, P.; Tan, Y.; Zhang, Z.; Li, J. An optimized EV charging model considering TOU price and SOC curve. *IEEE Trans. Smart Grid* **2012**, *3*, 388–393. [CrossRef]
19. Bessa, R.J.; Matos, M.A. Global against divided optimisation for the participation of an EV aggregator in the day-ahead electricity market. Part I: Theory. *Electr. Power Syst. Res.* **2013**, *95*, 309–318. [CrossRef]
20. Latinopoulos, C. Efficient Operation of Recharging Infrastructure for the Accommodation of Electric Vehicles: A Demand Driven Approach. Ph.D. Thesis, Imperial College London, London, UK, 2016. Available online: https://spiral.imperial.ac.uk/handle/10044/1/33340 (accessed on 25 December 2019).
21. Rotering, N.; Ilic, M.D. Optimal charge control of plug-in hybrid electric vehicles in deregulated electricity markets. *IEEE Trans. Power Syst.* **2011**, *26*, 1021–1029. [CrossRef]
22. Zhang, B.; Lam, A.Y.; Dominguez-Garcia, A.D.; Tse, D. An optimal and distributed method for voltage regulation in power distribution systems. *IEEE Trans. Power Syst.* **2015**, *30*, 1–13. [CrossRef]
23. Shokrzadeh, S.; Ribberink, H.; Rishmawi, I.; Entchev, E. A simplified control algorithm for utilities to utilize plug-in electric vehicles to reduce distribution transformer overloading. *Energy* **2017**, *133*, 1121–1131. [CrossRef]
24. Deforest, N.; Macdonald, J.S.; Black, D.R. Day ahead optimization of an electric vehicle fleet providing ancillary services in the Los Angeles Air Force Base vehicle-to-grid demonstration. *Appl. Energy* **2018**, *210*, 987–1001. [CrossRef]
25. Vandael, S.; Boucké, N.; Holvoet, T.; de Craemer, K.; Deconinck, G. Decentralized Coordination of Plug-In Hybrid Vehicles for Imbalance Reduction in a Smart Grid. In Proceedings of the AAMAS '11: The 10th International Conference on Autonomous Agents and Multiagent Systems, Taipei, Taiwan, 2–6 May 2011.
26. Yao, L.; Lim, W.H.; Tsai, T.S. A real-time charging scheme for demand response in electric vehicle parking station. *IEEE Trans. Smart Grid* **2017**, *8*, 52–62. [CrossRef]
27. Shafie-Khah, M.; Heydarian-Forushani, E.; Osorio, G.J.; Gil, F.A.S.; Aghaei, J.; Barani, M.; Catalao, J.P.S. Optimal behavior of electric vehicle parking lots as demand response aggregation agents. *IEEE Trans. Smart Grid* **2015**, *7*, 2654–2665. [CrossRef]
28. Zhang, G.; Tan, S.T.; Wang, G.G. Real-time smart charging of electric vehicles for demand charge reduction at non-residential sites. *IEEE Trans. Smart Grid* **2018**, *9*, 4027–4037. [CrossRef]
29. Su, W.; Chow, M.-Y. Performance evaluation of an EDA-based large-scale plug-in hybrid electric vehicle charging algorithm. *IEEE Trans. Smart Grid* **2011**, *3*, 308–315. [CrossRef]
30. Mehta, R.; Srinivasan, D.; Khambadkone, A.M.; Yang, J.; Trivedi, A. Smart charging strategies for optimal integration of plug-in electric vehicles within existing distribution system infrastructure. *IEEE Trans. Smart Grid* **2018**, *9*, 299–312. [CrossRef]
31. Babic, J.; Carvalho, A.; Ketter, W.; Podobnik, V. Extending Parking Lots with Electricity Trading Agent Functionalities. In Proceedings of the Workshop on Agent-Mediated Electronic Commerce and Trading Agent Design and Analysis (AMEC/TADA 2015), Istanbul, Turkey, 4 May 2015.
32. Moradijoz, M.; Moghaddam, M.P.; Haghifam, M.R.; Alishahi, E. A multi-objective optimization problem for allocating parking lots in a distribution network. *Int. J. Electr. Power Energy Syst.* **2013**, *46*, 115–122. [CrossRef]
33. Hashimoto, S.; Kanamori, R.; Ito, T. Auction-Based Parking Reservation System with Electricity Trading. In Proceedings of the 2013 IEEE 15th Conference on Business Informatics (CBI), Vienna, Austria, 15–18 July 2013; pp. 33–40.
34. Daina, N.; Sivakumar, A.; Polak, J.W. Electric vehicle charging choices: Modelling and implications for smart charging services. *Transp. Res. Part C Emerg. Technol.* **2017**, *81*, 36–56. [CrossRef]
35. Fazelpour, F.; Vafaeipour, M.; Rahbari, O.; Rosen, M.A. Intelligent optimization to integrate a plug-in hybrid electric vehicle smart parking lot with renewable energy resources and enhance grid characteristics. *Energy Convers. Manag.* **2014**, *77*, 250–261. [CrossRef]
36. Chung, Y.-W.; Khaki, B.; Li, T.; Chu, C.; Gadh, R. Ensemble machine learning-based algorithm for electric vehicle user behavior prediction. *Appl. Energy* **2019**, *254*, 113732. [CrossRef]
37. Latinopoulos, C.; Sivakumar, A.; Polak, J. Response of electric vehicle drivers to dynamic pricing of parking and charging services: Risky choice in early reservations. *Transp. Res. Part C Emerg. Technol.* **2017**, *80*, 175–189. [CrossRef]

38. van Ryzin, G.; McGill, G. Revenue management without forecasting or optimisation: An adaptive algorithm for determining airline seat protection levels. *Manag. Sci.* **2000**, *46*, 760–775. [CrossRef]
39. McGill, J.I.; van Ryzin, G.J. Revenue management: Research overview and prospects. *Transp. Sci.* **1999**, *33*, 233–256. [CrossRef]
40. Guadix, J.; Onieva, L.; Muñuzuri, J.; Cortés, P. An overview of revenue management in service industries: An application to car parks. *Serv. Ind. J.* **2011**, *31*, 91–105. [CrossRef]
41. Akhavan-Tabatabaei, R.; Bolívar, M.A.; Hincapie, J.A.; Medaglia, A.L. On the optimal parking lot subscription policy problem: A hybrid simulation-optimisation approach. *Ann. Oper. Res.* **2014**, *222*, 29–44. [CrossRef]
42. Teodorović, D.; Lučić, P. Intelligent parking systems. *Eur. J. Oper. Res.* **2006**, *175*, 1666–1681. [CrossRef]
43. Flath, C.M.; Gottwalt, S.; Ilg, J.P. A Revenue Management Approach for Efficient Electric Vehicle Charging Coordination. In Proceedings of the HICSS '12 Proceedings of the 45th Hawaii International Conference on System Sciences, Washington, DC, USA, 4–7 January 2012; pp. 1888–1896.
44. Ghosh, A.; Aggarwal, V. Control of Charging of Electric Vehicles through Menu-Based Pricing. In Proceedings of the IEEE International Conference on Communications (ICC), Paris, France, 21–25 May 2017.
45. Elexon. Imbalance Pricing Guidance. A Guide to Electricity Imbalance Pricing. 2019. Available online: www.elexon.co.uk/documents/training-guidance/bsc-guidance-notes/imbalance-pricing/ (accessed on 5 January 2020).
46. Vickrey, W.S. Congestion theory and transport investment. *Am. Econ. Rev.* **1969**, *59*, 251–260.
47. Daina, N. Modelling Electric Vehicle Use and Charging Behavior. Ph.D. Thesis, Imperial College London, London, UK, 2014. Available online: https://spiral.imperial.ac.uk:8443/handle/10044/1/25018 (accessed on 28 December 2019).
48. Hess, S.; Polak, J.W.; Daly, A.; Hyman, G. Flexible substitution patterns in models of mode and time of day choice: New evidence from the UK and The Netherlands. *Transportation* **2006**, *34*, 213–238. [CrossRef]
49. Gallego, G.; Iyengar, G.; Phillips, R.; Dubey, A. *Managing Flexible Products on a Network. Report TR-2004-01*; Computational Optimisation Research Center, Department of Industrial Engineering and Operations Research, Columbia University: New York, NY, USA, 2004.
50. Li, H.; Huh, W.T. Pricing multiple products with the multinomial logit and nested logit models: Concavity and implications. *Manuf. Serv. Oper. Manag.* **2011**, *13*, 549–563. [CrossRef]
51. Hetrakul, P.; Cirillo, C. A latent class choice based model system for railway optimal pricing and seat allocation. *Transp. Res. Part E Logist. Transp. Rev.* **2014**, *61*, 68–83. [CrossRef]
52. Battistelli, C.; Baringo, L.; Conejo, A. Optimal energy management of small electric energy systems including V2G facilities and renewable energy sources. *Electr. Power Syst. Res.* **2012**, *92*, 50–59. [CrossRef]
53. London Travel Demand Survey. 2011. Available online: www.clocs.org.uk/wp-content/uploads/2014/05/london-travel-demand-survey-2011.pdf (accessed on 4 January 2020).
54. United States Environmental Protection Agency (EPA) (2020) Fuel Economy. Available online: www.epa.gov/fueleconomy/ (accessed on 17 July 2020).
55. Yang, Z.; Li, K.; Foley, A. Computational scheduling methods for integrating plug-in electric vehicles with power systems: A review. *Renew. Sustain. Energy Rev.* **2015**, *51*, 396–416. [CrossRef]
56. Reid, D.J. Genetic algorithms in constrained optimisation. *Math. Comput. Model.* **1996**, *23*, 87–111. [CrossRef]
57. Aghajani, S.; Kalantar, M. A cooperative game theoretic analysis of electric vehicles parking lot in smart grid. *Energy* **2017**, *137*, 129–139. [CrossRef]
58. Hutson, C.; Venayagamoorthy, G.K.; Corzine, K.A. Intelligent Scheduling of Hybrid and Electric Vehicle Storage Capacity in a Parking Lot for Profit Maximization in Grid Power Transactions. In Proceedings of the IEEE Energy 2030 Conference, Atlanta, GA, USA, 17–18 November 2008; pp. 1–8.
59. Sehar, F.; Pipattanasomporn, M.; Rahman, S. Demand management to mitigate impacts of plug-in electric vehicle fast charge in buildings with renewables. *Energy* **2017**, *120*, 642–651. [CrossRef]
60. Alirezaei, M.; Noori, M.; Tatari, O. Getting to net zero energy building: Investigating the role of vehicle to home technology. *Energy Build.* **2016**, *130*, 465–476. [CrossRef]
61. Quddus, A.; Shahvari, O.; Marufuzzaman, M.; Usher, J.M.; Jaradat, R. A collaborative energy sharing optimization model among electric vehicle charging stations, commercial buildings, and power grid. *Appl. Energy* **2018**, *229*, 841–857. [CrossRef]

Article

Multi-Criteria Stochastic Selection of Electric Vehicles for the Sustainable Development of Local Government and State Administration Units in Poland

Paweł Ziemba

Faculty of Economics, Finance and Management, University of Szczecin, Cukrowa 8, 71-004 Szczecin, Poland; pawel.ziemba@usz.edu.pl

Received: 2 November 2020; Accepted: 27 November 2020; Published: 29 November 2020

Abstract: Increasing the popularity of electric vehicles is one way of reducing greenhouse gas emissions and making the economy more sustainable. In Poland, the use of electric vehicles is to be increased by the adoption of the Act on Electromobility and Alternative Fuels. This Act obliges local government units and state administration to expand the electric vehicle fleet. The expansion of the fleet should be carried out on a planned basis, based on rational decisions supported by economic analyses. Therefore, the aim of this article is to provide a recommendation of an electric vehicle that meets the needs of local and state administration to the greatest extent possible. The aim has been achieved using the multi-criteria decision analysis method called PROSA-C (PROMETHEE for Sustainability Assessment—Criteria) combined with the Monte Carlo method. The PROSA-C method allows promoting more sustainable vehicles with high technical, economic, environmental and social parameters. The Monte Carlo method, on the other hand, is a stochastic simulation tool that allows for taking into account the uncertainty of parameters describing vehicles. As a result of the research, the most and least attractive vehicles were identified from the perspective of the needs of local government units and state administration. Moreover, the conducted research allowed confirming the effectiveness and usefulness of the research methodology proposed in the article and the procedural approach combining the PROSA-C and Monte Carlo methods.

Keywords: electric vehicles; PROSA; PROMETHEE for Sustainability Assessment; MCDA; Multi-Criteria Decision Analysis; stochastic analysis; Monte Carlo; uncertainty

---

1. Introduction

Global energy demand is almost steadily increasing, the only exception being the global lockdown in the first half of 2020, which caused a temporary drop in demand for energy production [1]. Together with increased energy production, the number of greenhouse gases emitted into the atmosphere is also increasing: at present, more than 25 billion tons of CO_2 are released annually as a result of human activity [2]. One of the most important options for reducing CO_2 emissions into the atmosphere is the development of electric vehicles [3], because, according to the International Energy Agency, around 14% of greenhouse gases in the world are produced by the transport sector [4]. Therefore, the electrification of transport, together with an increase in the use of renewable energy sources, offers the possibility of significantly reducing greenhouse gas emissions [3]. Of course, the development of electric vehicles is not the only option for reducing CO_2, emissions associated with transport. Another interesting option is, for example, the design of urban space in such a way as to maximize the possibility of walking [5,6]. However, the need to use the vehicle cannot always be eliminated, so electrifying transport remains the main way to reduce emissions contributing to the sustainability of the transport system.

In order to increase the share of electric vehicles in transport, the Polish government has adopted the Act on Electromobility and Alternative Fuels [7] imposing obligations on state and local government units in terms of:

- the percentage of battery electric vehicles in the fleet of used vehicles,
- the percentage of zero-emission buses in the fleet of vehicles in use,
- the minimum number of charging points in municipal districts.

According to this legal act, state administration authorities and local government units with more than 50,000 inhabitants must gradually increase the share of electric vehicles in the vehicle fleet. For state administration bodies, the share of battery electric vehicles in the fleet of used vehicles should be: from 1 January 2022 at least 10%, from the beginning of 2023 at least 20%, and in 2025 at least 50%. For larger municipal districts and counties, i.e., with more than 50,000 inhabitants, the share should be at least 10% from 1 January 2022, and at least 30% in 2025. In addition, local government units are obliged to provide public transport services using a fleet partly composed of zero-emission buses: from the beginning of 2021 such buses are to constitute 5%, from 2023 10%, from 2025 20%, and on 1 January 2028 it is to be 30% of the bus fleet. As far as the expansion of the electric vehicle charging network is concerned, the number of charging points required in 2021 depends on the number of inhabitants and the number of cars in each municipal district. The detailed requirements are shown in Table 1 [7,8].

Table 1. Required number of electric vehicle charging points installed by 31 March 2021 depending on the size of the municipal district.

Number of Inhabitants of the Municipal District	Number of Motor Vehicles in the Municipal District	Number of Motor Vehicles per 1000 Inhabitants	Required Number of Charging Points
≥100,000	≥60,000	≥400	≥60
≥150,000	≥95,000	≥400	≥100
≥300,000	≥200,000	≥500	≥210
≥1,000,000	≥600,000	≥700	≥1000

Increasing the share of electric vehicles in the fleets of state and local government units is expected to contribute to the number of 1 million electric vehicles in use in Poland in 2025 [9]. Therefore, a critical problem is the appropriate expansion of the fleets of state and local administration vehicles. The expansion of fleets should be well thought-out and carried out in a planned manner, and decisions taken in this area must be rational and supported by analyses. The expansion of fleets of electric vehicles cannot take place chaotically, without a plan and research, because this can lead to wasteful spending of state and local government funds, with criminal sanctions associated with this. This is all the more important because the cost of purchasing electric vehicles is much higher than that of conventional vehicles, so ill-considered purchases of electric vehicles result in greater financial losses than conventional vehicles. Selecting suitable vehicles to expand the fleet is even more difficult due to the fact that electric vehicles, even belonging to the same class, differ in a number of parameters. It is important to properly select such vehicles in order to minimize the inconvenience associated with, among other things, the low range and long charging time of such vehicles, while maintaining a reasonable price of expanding the fleet of electric vehicles.

The research problem addressed in the article, and at the same time its practical objective, is to analyze the electric vehicles available in Poland and to present recommendations of vehicles with the highest utility for institutional users (state and local government bodies). This is of particular importance in the context of the previously indicated statutory obligation of state and local government units to expand the fleet of electric vehicles and the expected increase in demand for such vehicles.

The methodological contribution is the use of MCDA (Multi-Criteria Decision Analysis) method called PROSA-C (PROMETHEE for Sustainability Assessment–Criteria) [10,11], which allows taking into account the sustainability of decision-making alternatives in the solution of a decision problem. In the case of electric vehicles, the PROSA-C method allows promoting vehicles with more balanced economic, technical, social and environmental characteristics. What is important, in this article the method has been developed with elements of a stochastic simulation, which takes into account the uncertainty of individual parameters of electric vehicles. This uncertainty should be taken into account in the solution of the decision problem and is due, among other things, to the fact that certain operational parameters depend on the user and how the vehicle is used, e.g., the same vehicle may have different energy consumption and range depending on the driving style.

Section 2 contains a literature discussion on the applications of MCDA methods in decision-making problems related to electric vehicles. Section 3 discusses the PROSA-C method, a stochastic analysis using the Monte Carlo method, and a research procedure based on these methods. Section 4 presents the results of research on the selection of electric vehicles for local government units and state bodies in Poland. The article ends with a summary and indication of further research directions.

2. Literature Review

Electric vehicles are characterized by many parameters, such as: electric motor size, weight, range, battery capacity, energy consumption, charging time, etc. [12,13]. These parameters are often in conflict with each other, so improving one of them causes deterioration of another, e.g., higher battery capacity means longer charging time. Moreover, many parameters are characterized by uncertainty, e.g., charging time depends on the charger used. Therefore, the comparison and recommendation of electric vehicles is a multi-criteria problem characterized by uncertainty and consisting in the search for pareto-optimal solutions. MCDA methods are used to solve these types of decision-making problems [14]. These methods are used both to solve decision-making problems at the strategic level (e.g., selecting policies for the development of electric vehicles and their infrastructure) and at the tactical and operational level (e.g., selecting specific vehicles or locations of charging stations). Barfod et al. [15] studied the opportunities, risks and policies for the widespread use of commercial electric vehicles in Denmark. Adhikari et al. [16] examined the limitations and challenges of using electric vehicles in the context of Nepal. Both Zhang et al. [17] and Liu and Wei [18] examined the risk factors for building an electric vehicle charging infrastructure in a public-private partnership, with Zhang et al. [17] conducting the study from a Chinese perspective, and Liu and Wei [18] considered the three Chinese provinces separately, building their ranking. Fazeli et al. [19] assessed the potential impact of various fiscal policies on the acceptance of electric vehicles by consumers and the Government of Iceland over the next 30 years. Erbas et al. [20] and Ju et al. [21] examined the potential locations of electric vehicle charging stations and built a ranking, with Erbas et al. [20] conducting a survey for Ankara and Ju et al. [21] examining the locations in Beijing. Xu et al. [22] ranked the various electric vehicle sharing programs used in Beijing. Sałabun, Karczmarczyk [13] and Wątróbski et al. [23] compared different electric cars in their rankings. The decision-making problems considered in individual studies and the MCDA methods used are presented in Table 2.

In the above mentioned works, various MCDA methods were used, starting from methods based on single synthesizing criterion: AHP and Fuzzy AHP, TOPSIS and Fuzzy TOPSIS, VIKOR, SMARTER, DEMATEL, COMET, and ending with the outranking PROMETHEE II method. It should be noted here that the AHP method in each of the quoted tests was only used to determine the weights of the criteria. A detailed review of these and many other MCDA methods can be found, for instance, in the papers of [14,24]. In the context of the research objective, it should be noted that MCDA methods may have high or low compensation of criteria and therefore a correspondingly weak and strong sustainability. Depending on the adopted paradigm (strong or weak sustainability), different MCDA methods can be used to solve sustainability decision-making problems. Generally speaking, it is recognized that methods based on a single synthesizing criterion, such as AHP, SMARTER, TOPSIS, VIKOR, DEMATEL, COMET, among others, are characterized by high compensation and therefore only allow for weak sustainability. On the other hand, methods based on outranking, thanks to the use of indifference, preference and sometimes also veto thresholds, are characterized by lower compensation and thus stronger sustainability [25,26]. In the light of the purpose of the research, in addition to the degree of compensation, another important feature of MCDA methods is their ability to take into account the uncertainty of the criteria. The MCDA methods are often used to deal with *ex-ante* decision-making problems, so the decision-maker or analyst is not able to fully and certainly foresee all its consequences at the time of the decision. So this is a decision taken in conditions of uncertainty. A natural tool for taking into account uncertainty in decision making problems is fuzzy MCDA methods [27]. In turn, the MCDA methods using crisp data can take into account uncertainty by using a stochastic approach, such as the Monte Carlo method [28].

Table 2. Decision-making problems related to electric vehicles and MCDA methods used.

Decision Problem	Location	MCDA Method	No. of Alternatives/Criteria	Reference
Analysis of the challenges related to the widespread use of commercial electric vehicles	Denmark	SMARTER	-/25	[15]
Analysis of obstacles related to the use of electric vehicles	Nepal	AHP	-/17	[16]
Risk analysis of the development of electric vehicle charging infrastructure in public-private partnership	China	2-tuple DEMATEL	-/22	[17]
Risk assessment of the development of electric vehicle charging infrastructure in public-private partnership	China	Fuzzy TOPSIS	3/17	[18]
Assessment of the impact of fiscal policy on the acceptance of electric vehicles	Iceland	TOPSIS	6/4	[19]
Evaluation of the locations of electric vehicle charging stations	Ankara, Turkey	Fuzzy AHP + TOPSIS	12/15	[20]
Evaluation of the locations of electric vehicle charging stations	Beijing, China	Fuzzy AHP + PFWIG + GRP	6/14	[21]
Ranking of electric vehicle sharing programs	Beijing, China	VIKOR	3/22	[22]
Electric vehicle selection	Poland	COMET	9/6	[13]
Electric vehicle selection	Poland	PROMETHEE II, Fuzzy TOPSIS	36/9	[23]

Analyzing the quoted articles, several important research gaps can be observed in the studies on the selection of electric vehicles for public administration and local authorities. First of all, the presented literature review indicates a small number of works in which electric vehicles were compared in a multi-criteria way. It can also be noted that to the best of the author's knowledge, no scientific research on the selection or recommendation of electric vehicles for administrative and local government units has been published so far. On the other hand, the quoted research on the Polish electric vehicle market is out of date, because this market is changing dynamically and new vehicle models appear on it every year, and current vehicles are updated. As regards the MCDA methods used, it should be noted that among the approaches taking into account the uncertainty of parameters of electric vehicles, the only methods used are Fuzzy TOPSIS and COMET, operating on fuzzy sets. Unfortunately, both methods are characterized by high compensation [24] and therefore weak sustainability [14]. This means that a high value of one of the vehicle's parameters can fully compensate for a low value of another parameter [29], so, for example, a high maximum speed can compensate for high energy consumption. In other words, in the case of high compensation, a profit for one criterion may compensate for losses for another criterion, while low or no compensation reduces or eliminates this possibility altogether [30]. There is another research gap here, because the use of highly compensated MCDA methods results in solutions obtained using the indicated methods are not sustainable.

The research gaps identified directly determine the objectives and contribution of this research. The aim is a multi-criteria analysis of electric vehicles currently available in Poland, which meet the needs of state and local administration bodies. The practical effect will be to identify the most useful vehicles for these administrative units. As far as the scientific contribution is concerned, in this article, the PROSA-C method was used to evaluate electric vehicles available in Poland. It is a method designed to solve multi-criteria decision-making problems, enabling sustainable solutions. It allows adjusting the balance between the criteria influencing the expected degree of sustainability of the solution. This article develops the PROSA-C method by introducing elements of a stochastic analysis, thanks to which the uncertainty of data describing characteristics of electric vehicles is taken into account.

3. Materials and Methods

3.1. PROSA-C Method

PROSA is used to consider discrete decision-making problems, where a set of $A = \{a, b, \ldots\}$ consisting of m alternatives is considered. Alternatives are considered in terms of n criteria belonging to a set $C = \{c_1, c_2, \ldots, c_n\}$. The calculation procedure of the PROSA-C method [10,11] consists of 7 stages, with the initial 4 stages taken directly from the PROMETHEE method [31,32], on which the PROSA method is based. These steps are based on the approach using single criterion net flows [33] (pp.161–162) [31] (pp.200). The subsequent stages are written below using mathematical formulas.

1. Determination of deviations based on pair-wise comparisons:

$$d_j(a,b) = c_j(a) - c_j(b) \tag{1}$$

   where $d_j(a,b)$ denotes the difference between the performances of a and b on a j-th criterion.

2. Application of the preference function:

$$P_j(a,b) = F_j\left[d_j(a,b)\right] \tag{2}$$

   where $P_j(a,b)$ denotes the preference of alternative a with regard to alternative b on each criterion, as a function F of $d_j(a,b)$.

3. Calculation of a single criterion net outranking flow:

$$\phi_j(a) = \frac{1}{m-1}\sum_{i=1}^{m}\left[P_j(a,b_i) - P_j(b_i,a)\right] \tag{3}$$

   where $\phi_j(a)$ denotes the criterion net flow of a over any other alternative, and m describes the number of alternatives.

4. Calculation of a global (overall) net outranking flow:

$$\phi_{net}(a) = \sum_{j=1}^{n}\phi_j(a)\, w_j \tag{4}$$

where $\phi_{net}(a)$ is the weighted sum of net flow for each criterion, w_j is the weight of j-th criterion, and weights are normalized to 1 ($\sum_{j=1}^{n} w_j = 1$). The above steps allow obtaining the PROMETHEE II solution, while the subsequent steps lead to the PROSA-C solution.

5. Determination of a single criterion absolute deviation:

$$AD_j(a) = |\phi_{net}(a) - \phi_j(a)| s_j \quad (5)$$

where s_j denotes the sustainability (compensation) coefficient for a j-th criterion. As it can be easily noticed, s_j is a sort of a weight coefficient, and $AD_j(a)$ is a weighted distance of a solution $\phi_{net}(a)$ from solutions $\phi_j(a)$ obtained for individual criteria. The greater the value s_j, the more preferred are alternatives strongly sustainable with regard to the j-th criterion, therefore, the compensation degree for the criterion $c_j(a)$ is smaller.

6. Calculation of a single criterion PROSA net sustainable value:

$$PSV_j(a) = \phi_j(a) - AD_j(a) \quad (6)$$

where $PSV_j(a)$ describes the sustainability of alternative a and with regard to the j-th criterion.

7. Calculation of a global PROSA net sustainable value:

$$PSV_{net}(a) = \sum_{j=1}^{n} PSV_j(a)\, w_j \quad (7)$$

where $PSV_{net}(a)$ is the weighted sum of the PROSA net sustainable value for each criterion.

The PROSA-C method allows conducting an analysis of the sustainability of criteria for individual decision alternatives. It distinguishes three sustainability/compensation relationships.

1. The relation of being sustainable ($\approx$)–takes place when $\phi_j(a) \approx \phi_{net}(a)$ and it means that the alternative a is sustainable in terms of a j-th criterion.
2. The relation of being compensated (Cd)–takes place when $\phi_j(a) \lll \phi_{net}(a)$ and it means that the low performance of the criterion $c_j(a)$ is compensated by another criterion or other criteria ($\exists \phi_{j'}(a) : \phi_j(a)\ Cd\ \phi_{j'}(a)$).
3. The compensation relationship (Cs)–occurs when $\phi_j(a) \ggg \phi_{net}(a)$ and it means that the high performance of the criterion $c_j(a)$ compensates for lower performance on another criterion or other criteria ($\exists \phi_{j'}(a) : \phi_j(a)\ Cs\ \phi_{j'}(a)$).

Relations Cd and Cs are relations indicating the unsustainability of the alternative a with regard to the criterion $c_j(a)$. The << and >> operators denote contractual relations "much less than" and "much greater than", expressing the subjective views of the decision-maker about the difference between the compared values. These relations can provide a hint to the decision-maker about the expected values of the coefficient s_j. For example, if the decision-maker wants to increase the impact of sustainability on the obtained solution, a lower s_j value can be adopted for more sustainable criteria, and higher s_j values can be adopted for less sustainable criteria.

3.2. Stochastic Analysis

A stochastic analysis is based on random variables defined in the probability space. In this article, a stochastic simulation based on the Monte Carlo method is used. In the Monte Carlo method, K iterations are carried out, and during each of them L independent random variables r with a specific distribution D are drawn [34]:

$$r_l^k \overset{i.i.d.}{\sim} D(v) \quad (8)$$

where r_l^k is an l-th independent random variable (l = 1,2, ... ,L) drawn in a k-th iteration (k = 1,2, ... ,K), v is a set of distribution parameters D (e.g., range of values), *i.i.d.* means independent and identically distributed. This approach allows considering solutions based on different values of random variables, covering the whole probability space, including the specified distribution.

In this article, the Monte Carlo method has been used to generate various possible values for criteria describing uncertain parameters of electric vehicles. Random variables represent the values of individual criteria for each of the alternatives $c_j(a)$. Therefore, the number of random variables generated in each iteration should be equal to the product of the number of criteria and alternatives considered ($L = n * m$). However, each random variable may have a different distribution and distribution parameters, depending on which alternative and which criterion it represents. Therefore, in this article, the stochastic simulation is based on a formula:

$$c_j(a_i)^k = r_l^k \overset{i.i.d.}{\sim} D(v)_{i,j} \tag{9}$$

where $D(v)_{i,j}$ denotes the distribution and parameters adopted for a j-th criterion of an i-th alternative.

Further elements of the stochastic analysis are based on elements of SMAA (Stochastic Multicriteria Acceptability Analysis) [35]. On the basis of random variables generated in each iteration, rankings of alternatives are built using the PROSA-C method. The results obtained in subsequent iterations allow us to determine the statistic B_{ir} indicating the number of iterations, in which the i-th alternative obtained an r-th position in the ranking. On this basis, after K iterations, rank acceptability indices [35] are estimated:

$$b_i^r \approx \frac{B_{ir}}{K} * 100\% \tag{10}$$

In practice, the rank acceptability indices b_i^r show the probability that the i-th alternative will get the r-th position in the ranking. This probability is, of course, determined by random variables generated on the basis of the specified distributions $D(v)_{i,j}$. The obtained values of the rank acceptability indices are characterized by a certain precision and an assumed confidence interval, depending on the number of iterations. The expected precision of the rank acceptability indices can be determined from the formula [36]:

$$\varepsilon = \frac{z_\alpha}{\sqrt{4L}} \tag{11}$$

where z_α denotes a standard score for a confidence level $1 - \alpha$. In the SMAA method and Monte Carlo simulations 10^4–10^6 iterations [36,37] are usually used, which gives a precision of 0.01–0.001 for 95% confidence respectively.

3.3. Research Procedure

Some elements of the research procedure have already been mentioned in Section 3.2, while this section discusses the complete procedure applied to the decision-making problem of recommending electric vehicles for state and local authorities. A diagram of the procedure is presented in a graphical form in Figure 1.

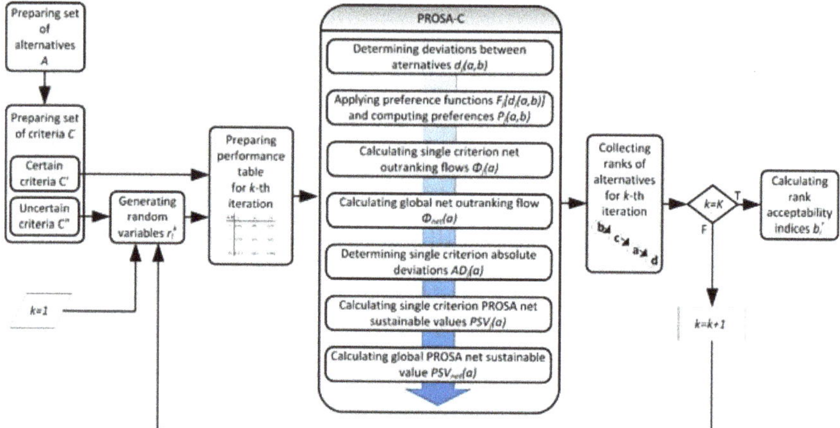

Figure 1. Research procedure.

1. At the beginning, an analysis of electric vehicles available in Poland was carried out and a review of sources describing usual needs of the state administration and local government in relation to electric vehicles was conducted. Data on electric vehicles available on the Polish market were obtained from the database of the Polish Alternative Fuels Association [38]. On the other hand, the needs of administration units with regard to electric vehicles were determined on the basis of information on tenders for such vehicles announced by administration units and information on purchases made by local government units and public administration units [39–42]. This information was also verified by consulting local government experts. On this basis, vehicles applicable in local government units and state authorities were selected, obtaining a set of decision-making alternatives A.
2. The vehicle characteristics were then considered and a set of criteria C was obtained. The set of criteria was established on the basis of the data on electric vehicles obtained in step 1, with the reservation that these must be measurable and, at least to some extent, objective criteria. For this reason, criteria such as the appearance of the vehicle were not taken into account. The selection of the set of criteria was also based on other studies on vehicle recommendations, including electric vehicles [12,13,23]. The following criteria were taken from these publications: Top speed [23], Power [13,23], Torque [13,23], Cargo capacity [23], Battery capacity [12,13,23], Price [13,23], Range [12,13,23], Consumption [12], Charging time [13,23], Fast charging time [23]. This collection was supplemented with the criteria obtained in an expert interview: Acceleration, Seats, and Safety. This set of criteria was divided into certain criteria, the values of which for each alternative are crisp numbers, and uncertain criteria, the values of which may vary according to different factors. In addition, ranges of uncertain criteria were established.
3. In the next stage, based on the Monte Carlo simulation, random variables were generated to determine the values of uncertain criteria.
4. A performance table with values for the certain criteria and simulated crisp values for the uncertain criteria was constructed.
5. On the basis of the performance table, calculations were carried out using the PROSA-C method, thereby obtaining a ranking of alternatives.
6. The procedure of generating random numbers and building the ranking of alternatives was repeated K-times (the study assumed the number K = 1,000,000 iterations).
7. After all iterations of the calculation procedure were performed, the rank acceptability indices b_i^r were calculated for each alternative.

4. Results

At the outset, the most frequent needs of state and local government units with regard to electric vehicles were identified. The analysis of Internet sources [39–42] as well as the expert interview revealed that both the state administration and local government in Poland most often use electric hatchback, B-segment (small cars), C-segment (medium cars) and J-segment (SUVs) cars with appropriate electric motor power, energy consumption and range. Therefore, electric cars of these segments with electric motor power of not less than 130PS, a range of at least 200 km and energy consumption of not more than 250 Wh/km are considered in this research. Moreover, only electric vehicles available on the Polish market were considered [38]. The vehicles concerned and their characteristics are shown in Table 3.

Table 3. Characteristics of the electric vehicles concerned [38,43,44].

	Parameter	A1-DS 3 CROSSBACK E-TENSE	A2-Hyundai KONA Electric 39.2 kWh	A3-Hyundai KONA Electric 64 kWh	A4-Hyundai IONIQ Electric	A5-Nissan LEAF	A6-Nissan LEAF e+	A7-BMW i3	A8-BMW i3s	A9-Mini Cooper SE	A10-Opel Corsa-e	A11-Peugeot e-208	A12-Renault ZOE R135
	Segment	JB	JB	JB	C	C	C	B	B	B	B	B	B
	Top speed [km/h]	150	155	167	165	144	157	150	160	150	150	150	140
	Acceleration 0–100 km/h [s]	8.7	9.9	7.9	9.7	7.9	7.3	7.3	6.9	7.3	8.1	8.1	9.5
	Total power [PS]	136	136	204	136	147	218	170	184	184	136	136	136
	Total torque [Nm]	260	395	395	295	320	340	250	270	270	260	260	245
	Cargo volume [L]	350	361	361	357	435	420	260	260	211	309	265	338
	Cargo volume–seats folded [L]	1050	1143	1143	1417	1176	1161	1100	1100	731	1118	1106	1225
	Seats [people]	5	5	5	5	5	5	4	4	4	5	5	5
	Battery capacity-useable [kWh]	45	39.2	64	38.3	36	56	37.9	37.9	28.9	45	45	52
Price [PLN th.]	Min equipment	159.9	158.4	182.4	184.5	155.5	164	169.7	184.2	139.2	128.9	124.9	135.9
	Max equipment	199.9	185.3	232.6	204	195.5	211.1	233.7	242.6	173.4	171.9	155.7	165.6
Range [km]	EVDB range	250	255	400	250	220	325	235	230	185	275	275	310
	WLTP range	320	305	484	311	270	385	308	283	234	330	339	385
	City–cold weather	250	250	390	235	215	320	235	230	180	270	270	305
	City–mild weather	375	385	595	365	325	485	365	355	280	410	415	465
	Highway–cold weather	175	180	280	175	155	230	165	160	130	195	195	220
	Highway–mild weather	225	230	365	230	200	300	215	205	170	250	255	280
	Combined–cold weather	210	215	335	205	185	275	200	195	155	230	230	260
	Combined–mild weather	285	295	460	290	250	375	275	265	215	315	320	355
Energy consumption [Wh/km]	EVDB vehicle consumption	180	154	160	153	164	172	161	165	156	164	164	168
	WLTP rated consumption	176	143	147	138	206	180	153	161	152	170	164	177
	WLTP vehicle consumption	141	129	132	123	133	145	123	134	124	136	133	135
	City–cold weather	180	157	164	163	167	175	161	165	161	167	167	170
	City–mild weather	120	102	108	105	111	115	104	107	103	110	108	112
	Highway–cold weather	257	218	229	219	232	243	230	237	222	231	231	236
	Highway–mild weather	200	170	175	167	180	187	176	185	170	180	176	186
	Combined–cold weather	214	182	191	187	195	204	190	194	186	196	196	200
	Combined–mild weather	158	133	139	132	144	149	138	143	134	143	141	146
Charging time [m]	Wall plug (2.3 kW)	1395	1215	1965	1185	1110	1725	1170	1170	900	1395	1395	1605
	1-phase 16A (3.7 kW)	870	750	1230	735	690	1080	735	735	555	870	870	1005
	1-phase 32A (7.4 kW)	435	375	615	375	390 [5]	600 [2]	375	375	285	435	435	510
	3-phase 16A (11 kW)	300 [6]	255	420	735 [1]	690	1080 [1]	255	255	195	300 [6]	300 [6]	345
	3-phase 32A (22 kW)	300 [6]	255 [4]	420 [4]	375 [3]	390 [5]	600 [2]	255 [4]	255 [4]	195 [4]	300 [6]	300 [6]	180

Table 3. *Cont.*

	Parameter	A1-DS 3 CROSSBACK E-TENSE	A2-Hyundai KONA Electric 39.2 kWh	A3-Hyundai KONA Electric 64 kWh	A4-Hyundai IONIQ Electric	A5-Nissan LEAF	A6-Nissan LEAF e+	A7-BMW i3	A8-BMW i3s	A9-Mini Cooper SE	A10-Opel Corsa-e	A11-Peugeot e-208	A12-Renault ZOE R135
Fast charging time [m]	CHAdeMO 50 kW DC					40	62						
	CCS 50 kW DC	50	50	63	50 [7]			36	36	29	50	50	56 [9]
	CHAdeMO 100 kW DC						35						
	CCS 175 kW DC	27 [12]	50 [10]	44 [11]	47 [8]			36 [10]	36 [10]	29 [10]	27 [12]	27 [12]	56 [9]
	CCS 350 kW DC	27 [12]	50 [10]	44 [11]	47 [8]			36 [10]	36 [10]	29 [10]	27 [12]	27 [12]	56 [9]
Safety [%]	Adult occupant	87	87	87	91	93	93	86	86	79	79	91	89
	Child occupant	86	85	85	80	86	86	81	81	73	77	86	80
	Pedestrian	54	62	62	70	71	71	57	57	66	71	56	66
	Safety assist	63	60	60	82	71	71	55	55	56	56	71	85

[1] 3.7 kW max; [2] 6.6 kW max; [3] 7.4 kW max; [4] 11 kW max; [5] optional 6.6 kW On-board Charger (3.6 kW is standard); [6] optional 11 kW On-board Charger (7.4 kW is standard); [7] 40 kW max; [8] 44 kW max; [9] 46 kW max; [10] 50 kW max; [11] 77 kW max; [12] 100 kW max.

On the basis of Table 3, the decision-making criteria were defined, which were: C1–Top Speed, C2–Acceleration, C3–Total Power, C4–Total Torque, C5–Cargo Volume, C6–Cargo Volume–Seats Folded, C7–Seats, C8–Battery Capacity, C9–Price, C10–Range, C11–Energy Consumption, C12–Charging Time, C13–Fast Charging Time, C14–Safety. It is easy to notice that criteria C9–C14 are uncertain and their values may vary depending on the situation. The other criteria (C1–C8) are certain and their values remain constant. Based on Table 3, the ranges of uncertain criteria values are defined in Table 4. Table 4 therefore shows the distribution parameters $D(v)_{i,j}$ used in the Monte Carlo simulation.

In subsequent iterations of the procedure, new values for uncertain criteria were generated on the basis of Table 4, using continuous uniform distribution for each criterion. These values, together with the values of the certain criteria, were recorded in the performance table and constituted the input for the PROSA-C method. In the PROSA-C method, in each iteration the same preference model was used, specifying the functions and directions of preferences and the weights of criteria. The weights may be defined directly or relative to other criteria [45], with the PROSA-C method expressing them directly. In the preference model, the most difficult to build is to select the preference function and set the values of the indifference and preference thresholds [46,47]. It is assumed that for quantitative criteria one of the functions should be used: V-shape criterion, V-shape criterion with indifference area, or Gaussian criterion. For qualitative criteria, the Usual criterion or Level criterion is most often used [48]. As regards the values of indifference (q) and preference (p) thresholds used in the V-shaped functions, Roy states that the threshold values should be between reliable minimum and maximum values for a given criterion. He also suggests that the values of these thresholds may be based on certain characteristics of the value of the criteria, such as mean value, standard deviation, minimum, maximum, etc. [49]. Deshmukh, on the other hand, notes in the context of the Gaussian criterion that the Gaussian threshold (σ) should be placed between the values of thresholds q and p [48]. Amponsah et al. interpret the threshold σ as the standard deviation of the value of a given criterion [50]. Based on these observations, it can be assumed that the threshold q should be less than the standard deviation of the criterion value and the threshold p should be greater than the standard deviation. In addition, Podvezko and Podvezko propose an approach where the values of thresholds q

and p should be between a minimum and a maximum value for the gap between the values of the alternatives on a given criterion [51,52]. The last parameter of the preference model in the PROSA-C method is the values of sustainability coefficients. Research on the PROSA-C method has shown that in order to maintain the performance scale [−1.1] in accordance with the classic PROMETHEE II method for a problem consisting of 14 criteria, sustainability coefficients with values no greater than 0.3 should be used [10]. Taking into account the above observations, a preference model presented in Table 5 was defined. The weights and functions of preferences were expertly defined taking into account assumptions about the applicability of the preferences function [48]. The preference thresholds were based on the population standard deviation σ_j calculated from all the values of a given criterion (for certain criteria) and all the minimum values of a given criterion (for uncertain criteria). The indifference threshold was set at $q = 0.5\sigma_j$ whereas the preference threshold at $p = 2\sigma_j$ [53]. Such threshold values are in line with previous considerations, and at the same time meet the request of Podviezko and Podviezko for threshold values to be included between a minimum and a maximum range of alternatives for a given criterion. Moreover, based on the analysis in [10], the values of sustainability coefficients were set as $s_j = 0.3$ for each criterion.

Table 4. Ranges of uncertain criteria.

Criterion		A1-DS 3 CROSSBACK E-TENSE	A2-Hyundai KONA Electric 39.2 kWh	A3-Hyundai KONA Electric 64 kWh	A4-Hyundai IONIQ Electric	A5-Nissan LEAF	A6-Nissan LEAF e+	A7-BMW i3	A8-BMW i3s	A9-Mini Cooper SE	A10-Opel Corsa-e	A11-Peugeot e-208	A12-Renault ZOE R135
C9–Price [PLN thousand]	min	159.9	158.4	182.4	184.5	155.5	164	169.7	184.2	139.2	128.9	124.9	135.9
	max	199.9	185.3	232.6	204	195.5	211.1	233.7	242.6	173.4	171.9	155.7	165.6
C10–Range [km]	min	175	180	280	175	155	230	165	160	130	195	195	220
	max	375	385	595	365	325	485	365	355	280	410	415	465
C11–Energy Consumption [Wh/km]	min	120	102	108	105	111	115	104	107	103	110	108	112
	max	257	218	229	219	232	243	230	237	222	231	231	236
C12–Charging Time [m]	min	300	255	420	375	390	600	255	255	195	300	300	180
	max	1395	1215	1965	1185	1110	1725	1170	1170	900	1395	1395	1605
C13–Fast Charging Time [m]	min	27	50	44	47	40	35	36	36	29	27	27	56
	max	50	50	63	50	40	62	36	36	29	50	50	56
C14–Safety [%]	min	54	60	60	70	71	71	55	55	56	56	56	66
	max	87	87	87	91	93	93	86	86	79	79	91	89

Table 5. Preference model.

Criterion	Weight	Preference Direction	Preference Function	Indifference Threshold (q)	Preference Threshold (p)
C1–Top Speed [km/h]	3	Max	5	4	16
C2–Acceleration 0–100 km/h [s]	2	Min	5	0.5	2
C3–Total Power [PS]	1	Max	5	15	61
C4–Total Torque [Nm]	1	Max	5	27	108
C5–Cargo Volume [L]	3	Max	5	34	136
C6–Cargo Volume -Seats Folded [L]	3	Max	5	77	309
C7–Seats [people]	5	Max	1	-	-
C8–Battery Capacity [kWh]	4	Max	3	-	19.4
C9–Price [PLN thousand]	5	Min	3	-	42.4
C10–Range [km]	5	Max	3	-	80

Table 5. *Cont.*

Criterion	Weight	Preference Direction	Preference Function	Indifference Threshold (q)	Preference Threshold (p)
C11–Energy Consumption [Wh/km]	3	Min	5	3	10
C12–Charging Time [m]	4	Min	5	57	230
C13–Fast Charging Time [m]	5	Min	3	-	20
C14–Safety [%]	5	Max	3	-	14

Preference function: 5–V-shape with indifference, 1–Usual criterion, 3–V-shape criterion.

Based on the conducted simulation, the rank acceptability indices b_i^r contained in Table 6 were calculated. Additionally, the values of the rank acceptability indices are graphically shown in Figure 2. One million iterations allowed obtaining accuracy of results at 0.1% with 95% confidence level.

Table 6. Rank acceptability indices for individual alternatives.

Rank	Rank Acceptability Index [%]											
	A1	A2	A3	A4	A5	A6	A7	A8	A9	A10	A11	A12
1 (b_i^1)	3.6	2.3	15.1	3.0	6.7	37.4	0	0	0	10.3	18.0	3.4
2 (b_i^2)	5.7	5.4	14.8	6.9	11.6	19.1	0	0	0	13.4	16.5	6.6
3 (b_i^3)	6.9	7.7	13.4	9.9	13.7	12.9	0.1	0.1	0	13.3	13.5	8.4
4 (b_i^4)	7.9	9.6	12.3	12.2	14.3	9.5	0.4	0.4	0	12.5	11.4	9.6
5 (b_i^5)	8.9	11.1	11.2	13.8	13.8	7.2	0.8	0.8	0	11.7	10.0	10.6
6 (b_i^6)	10.2	12.4	10.2	14.5	12.8	5.4	1.6	1.5	0	10.7	8.8	11.8
7 (b_i^7)	11.7	13.6	8.8	14.4	11.0	3.9	3.0	3.0	0.1	9.8	7.7	13.0
8 (b_i^8)	13.6	14.6	7.1	12.7	8.6	2.6	5.6	5.6	0.6	8.5	6.5	13.9
9 (b_i^9)	15.7	13.6	4.8	8.8	5.3	1.5	11.6	11.4	2.7	6.4	4.9	13.3
10 (b_i^{10})	9.7	6.9	1.8	3.1	1.7	0.4	26.3	26.1	12.6	2.6	2.1	6.7
11 (b_i^{11})	4.5	2.3	0.4	0.7	0.3	0.1	28.3	28.8	31.0	0.7	0.6	2.3
12 (b_i^{12})	1.5	0.5	0	0.1	0	0	22.2	22.2	52.9	0.1	0.1	0.4

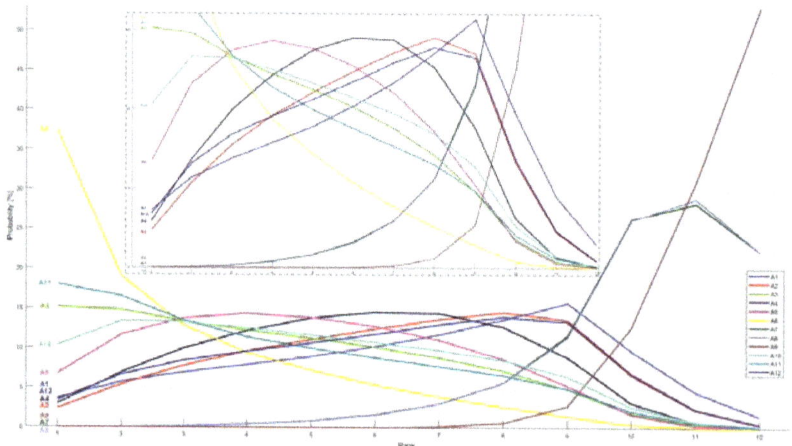

Figure 2. Plot of rank acceptability indices for individual alternatives A1–A12 (and zoomed fragment of the plot).

The analysis of Table 6 and Figure 2 shows that the best alternative, which most often comes first in the simulations, is A6, i.e., Nissan LEAF e+. Slightly worse results are obtained by the group of alternatives A11 (Peugeot e-208), A3 (Hyundai KONA Electric 64 kWh), A10 (Opel Corsa-e), but they are also in the lead and their profiles, determined by the ranks obtained in the simulations, are similar. The rankings are usually followed by the A5 (Nissan LEAF) and A4 (Hyundai IONIQ Electric) alternatives, which are followed by a group of alternatives with similar ranks: A1 (DS 3 CROSSBACK E-TENSE), A2 (Hyundai KONA Electric 39.2 kWh), A12 (Renault ZOE R135). The last positions in the rankings are usually very similar to A7 (BMW i3), A8 (BMW i3s) and A9 (Mini Cooper SE) alternatives, with Mini Cooper SE achieving the worst rank in over half of the simulation. The results obtained are, of course, still uncertain, i.e., it is impossible to determine the order of alternatives with unknown values for uncertain criteria. However, this order is largely predictable. Furthermore, it is possible to identify a group of vehicles whose purchase is worthwhile and a group of vehicles which should not be taken into account. The first group certainly includes A6—Nissan LEAF e+, and the following vehicles can also be considered: A3—Hyundai KONA Electric 64 kWh, A10—Opel Corsa-e and A11—Peugeot e-208. The second group should certainly include the following: A1—DS 3 CROSSBACK E-TENSE, A2—Hyundai KONA Electric 39.2 kWh, A12—Renault ZOE R135, A7—BMW i3, A8—BMW i3s, A9—Mini Cooper SE.

5. Discussion

The high rankings achieved by alternatives A3 and A6 are due to their good performance in terms of numerous criteria. The values of the criteria for the individual alternatives are shown in Figure 3. Bearing in mind that the preference direction for C1, C3–C8, C10, C14 is maximum and for C2, C9, C11–C13 the minimum is preferred, it is easy to observe that alternatives A3 and A6 have relatively high values of criteria C1–C8, C10, C14. The results are worse for criteria C12–C13 and values for the other criteria are average for the alternatives discussed. This allows alternatives A3 and A6 to obtain high values for ϕ_{net} (compare formula 4). Moreover, these alternatives are relatively balanced on criteria (C12–C13 is an exception). This results in low absolute deviation values (compare formula 5), so they do not receive large penalties for not reaching a sustainable balance of criteria. As a result, they have high PSV values.

As far as the highly rated alternatives A10 and A11 are concerned, they also score well on criteria C7, C9-C10, C13 and, for most of the other criteria, do not deviate significantly from the average. This gives them quite high scores for ϕ_{net}, but they gain higher profits in the PROSA-C method because they are more sustainable than alternatives A3 and A6. This observation can be confirmed by comparing the ranking of acceptability indices obtained in the PROSA-C method (Table 6) with the values obtained with the same assumptions (1 million iterations, accuracy 0.1%, 95% confidence level) in the classic PROMETHEE II method, presented in Table 7.

Comparing Tables 6 and 7, it is easy to see that in the case of alternatives A3 and A6, the PROSA-C sustainability study makes them ranked first less frequently compared to PROMETHEE, while alternatives A10 and A11 are more likely to win the PROSA-C rankings than in PROMETHEE II. Tables 8 and 9 show the rank acceptability indices obtained in the PROSA-C method with sustainability coefficients values of $s_j = 0.5$ and $s_j = 1$ respectively. The aim of this study was to check whether the results would change as the impact of the sustainability of alternatives on the achieved solution increased.

The analysis of Tables 6–9 shows that as the value of the sustainability coefficient increases, the alternatives A10 and A11 as well as A1, A2, A5 gain. In turn, the alternatives A3, A4, A6, A7, A8, A9, A12 lose. This comparison therefore allows dividing the alternatives into two groups, which include more and less sustainable alternatives.

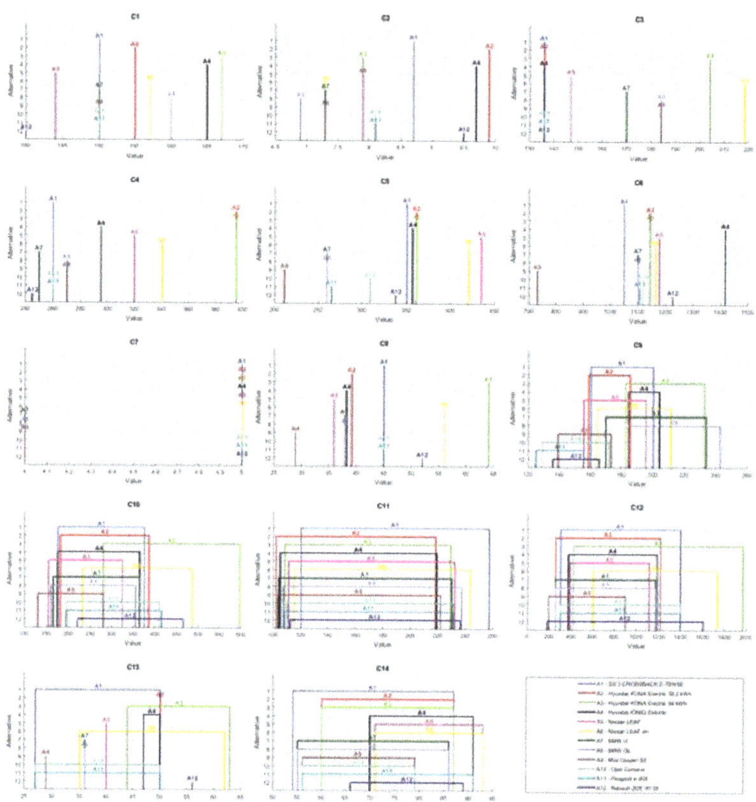

Figure 3. Values of the criteria C1-C14 for the different alternatives A1–A12.

Table 7. Rank acceptability indices for individual alternatives obtained by the PROMETHEE II method.

Rank	Rank Acceptability Index [%]											
	A1	A2	A3	A4	A5	A6	A7	A8	A9	A10	A11	A12
1 (b_i^1)	1.9	1.6	22.8	3.7	4.3	37.5	0.0	0.0	0.0	7.2	16.0	5.0
2 (b_i^2)	3.6	3.9	20.3	8.0	8.9	21.3	0.1	0.1	0.0	10.4	15.3	8.2
3 (b_i^3)	5.0	6.0	16.2	11.2	12.2	13.9	0.2	0.3	0.0	11.6	13.4	10.0
4 (b_i^4)	6.3	8.0	12.7	13.2	14.1	9.6	0.5	0.6	0.0	12.1	11.7	11.1
5 (b_i^5)	7.7	9.8	9.9	14.2	14.8	6.8	0.9	1.2	0.1	12.2	10.5	11.8
6 (b_i^6)	9.4	11.9	7.5	14.4	14.3	4.7	1.7	2.1	0.3	11.9	9.4	12.3
7 (b_i^7)	11.5	13.8	5.3	13.4	12.7	3.1	3.1	3.8	1.0	11.3	8.3	12.6
8 (b_i^8)	14.1	15.6	3.2	11.0	9.9	1.9	5.7	6.8	2.8	10.2	6.9	12.0
9 (b_i^9)	16.7	14.7	1.5	7.2	5.9	0.9	10.7	12.4	8.0	7.6	5.0	9.4
10 (b_i^{10})	12.3	9.0	0.4	2.9	2.2	0.2	19.6	21.9	20.4	3.7	2.4	5.0
11 (b_i^{11})	7.6	4.3	0.1	0.8	0.6	0.0	24.7	24.7	32.9	1.4	0.9	2.0
12 (b_i^{12})	3.9	1.5	0.0	0.1	0.1	0.0	32.8	26.1	34.4	0.4	0.2	0.5

Table 8. Rank acceptability indices for individual alternatives obtained by the PROSA-C method, $s_j = 0.5$.

Rank	Rank Acceptability Index [%]											
	A1	A2	A3	A4	A5	A6	A7	A8	A9	A10	A11	A12
1 (b_i^1)	5.0	2.9	11.6	2.4	8.3	36.1	0.0	0.0	0.0	12.5	18.7	2.5
2 (b_i^2)	7.2	6.4	11.9	6.1	13.0	18.1	0.0	0.0	0.0	14.9	16.9	5.6
3 (b_i^3)	8.1	9.0	11.3	9.2	14.4	12.5	0.1	0.1	0.0	14.0	13.7	7.5
4 (b_i^4)	9.0	10.7	11.2	11.7	14.4	9.5	0.4	0.3	0.0	12.6	11.4	8.8
5 (b_i^5)	9.7	11.9	11.0	13.6	13.4	7.5	0.9	0.6	0.0	11.4	9.9	10.1
6 (b_i^6)	10.7	12.8	10.8	14.7	12.0	5.9	1.7	1.3	0.0	10.1	8.6	11.5
7 (b_i^7)	11.7	13.4	10.4	14.9	10.2	4.5	3.1	2.7	0.0	8.8	7.4	12.9
8 (b_i^8)	12.9	13.5	9.6	13.6	7.8	3.2	5.9	5.2	0.2	7.4	6.2	14.6
9 (b_i^9)	13.8	11.9	7.8	9.7	4.8	1.9	12.3	10.8	1.2	5.5	4.7	15.5
10 (b_i^{10})	8.0	5.6	3.5	3.4	1.5	0.6	30.0	27.2	8.1	2.1	1.9	8.1
11 (b_i^{11})	3.3	1.6	1.0	0.7	0.2	0.1	30.0	33.1	26.3	0.5	0.5	2.7
12 (b_i^{12})	0.7	0.2	0.1	0.0	0.0	0.0	15.6	18.7	64.2	0.0	0.1	0.3

Table 9. Rank acceptability indices for individual alternatives obtained by the PROSA-C method, $s_j = 1$.

Rank	Rank Acceptability Index [%]											
	A1	A2	A3	A4	A5	A6	A7	A8	A9	A10	A11	A12
1 (b_i^1)	9.0	4.2	6.1	1.3	11.6	31.4	0.0	0.0	0.0	17.0	18.6	0.8
2 (b_i^2)	10.5	9.1	7.2	4.3	14.9	16.2	0.0	0.0	0.0	17.4	17.2	3.2
3 (b_i^3)	10.7	11.7	7.6	7.7	15.1	12.0	0.1	0.0	0.0	14.9	14.3	5.8
4 (b_i^4)	11.0	13.1	8.2	10.7	14.3	9.9	0.5	0.1	0.0	12.8	12.1	7.4
5 (b_i^5)	11.2	13.7	9.1	13.2	12.8	8.4	1.2	0.5	0.0	10.8	10.2	8.9
6 (b_i^6)	11.3	13.5	10.0	15.1	11.0	7.0	2.3	1.2	0.0	9.0	8.7	10.8
7 (b_i^7)	11.0	12.4	11.3	16.3	8.8	5.8	4.2	2.5	0.0	7.3	7.2	13.2
8 (b_i^8)	10.3	10.7	12.6	15.3	6.5	4.6	8.0	5.2	0.0	5.5	5.7	15.8
9 (b_i^9)	8.5	7.6	13.5	10.8	3.6	3.1	16.1	11.2	0.3	3.5	3.9	17.9
10 (b_i^{10})	4.6	3.2	9.4	4.4	1.1	1.3	32.9	25.6	2.7	1.4	1.7	11.6
11 (b_i^{11})	1.7	0.8	4.6	0.9	0.2	0.3	29.2	42.4	14.8	0.3	0.4	4.4
12 (b_i^{12})	0.1	0.0	0.5	0.0	0.0	0.0	5.6	11.2	82.2	0.0	0.0	0.3

6. Conclusions

The article considers the problem of analyzing and recommending electric vehicles which are most useful for local government and state administration units. For this purpose, based on publicly available information about tenders for local governments and state administration, the most typical demand of such units for electric vehicles was determined. Next, the vehicles available on the Polish market were analyzed, indicating vehicles meeting the needs of local and state units. Based on the relatively new MCDA method called PROSA-C and on the Monte Carlo method, the most useful vehicles were identified, taking into account the uncertainty of parameters describing individual vehicles. The conducted research indicated that the best choice, based on a defined preference model, is currently Nissan LEAF e+.

As regards the generalized conclusions of the studies carried out, there is a wide range of electric vehicles in B, C and J segments. Twelve such vehicles were included in the studies carried out, but a few others were initially rejected on the grounds that they did not meet the requirements for electric motor power or range. In addition, it is easy to see that many of the parameters characterizing electric vehicles are uncertain and dependent on the use of the vehicles or on external factors such as consumption or the types of charging station available. In relation to the methodological approach used in the studies, its usefulness is obviously not limited to electric vehicles, but the methodology is much more universal. The methodological contribution of the article consists in developing the PROSA method by a stochastic analysis based on the Monte Carlo method. This allows the PROSA method, naturally designed to work on crisp data, to take into account the uncertainty of the input data. In addition, the PROSA method, indicating the most useful solution, takes into account the sustainability of the alternatives considered and through the option of changing the values of the sustainability coefficients s_j allows seeking more or less sustainable solutions.

The limitations of the studies presented in the article are mainly related to the fact that one preference model, presented in Table 5, has been applied. Therefore, the study presented captures the uncertainty of the parameters of electric vehicles, but does not take into account the possible uncertainty of the weights of the criteria, as well as the possibility of applying other preference functions and other values of the indifference and preference thresholds. Another limitation refers to the constant changes on the Polish electric vehicle market. These changes are related, among other things, to the development of the electric vehicle market and the introduction of new vehicle models on it, and to changes in the legal environment of that market. These restrictions result in further directions of research.

An important methodological research direction will be to develop the PROSA method into Fuzzy PROSA, capable of capturing uncertainty by means of fuzzy numbers, as other MCDA fuzzy methods do [54–56]. As far as practical aspects are concerned, it should be noted that subsidies for individual users have been introduced in Poland for the purchase of electric vehicles [57,58]. This may significantly change the situation on the vehicle market and make electric vehicles more attractive and accessible to the public. It would be interesting to carry out a study to address this situation and consider the attractiveness of electric and hybrid vehicles compared to conventional (combustion) vehicles for individual users.

Funding: This work was supported by the National Science Centre, Poland, Grant no. 2019/35/D/HS4/02466.

Conflicts of Interest: The author declares no conflict of interest.

References

1. International Energy Agency Global Energy Review 2020 Analysis. Available online: https://www.iea.org/reports/global-energy-review-2020 (accessed on 28 November 2020).
2. Li, W.; Lin, Z.; Cai, K.; Zhou, H.; Yan, G. Multi-Objective Optimal Charging Control of Plug-In Hybrid Electric Vehicles in Power Distribution Systems. *Energies* **2019**, *12*, 2563. [CrossRef]
3. Taljegard, M.; Göransson, L.; Odenberger, M.; Johnsson, F. Electric Vehicles as Flexibility Management Strategy for the Electricity System—A Comparison between Different Regions of Europe. *Energies* **2019**, *12*, 2597. [CrossRef]
4. Park, E.; Lim, J.; Cho, Y. Understanding the Emergence and Social Acceptance of Electric Vehicles as Next-Generation Models for the Automobile Industry. *Sustainability* **2018**, *10*, 662. [CrossRef]
5. Abastante, F.; Lami, I.M.; La Riccia, L.; Gaballo, M. Supporting Resilient Urban Planning through Walkability Assessment. *Sustainability* **2020**, *12*, 8131. [CrossRef]
6. Dörrzapf, L.; Kovács-Győri, A.; Resch, B.; Zeile, P. Defining and assessing walkability: A concept for an integrated approach using surveys, biosensors and geospatial analysis. *Urban Dev. Issues* **2019**, *62*, 5–15. [CrossRef]
7. Ministry of Energy. *Ustawa o elektromobilności i paliwach alternatywnych*; 2018. Available online: http://prawo.sejm.gov.pl/isap.nsf/download.xsp/WDU20180000317/U/D20180317Lj.pdf (accessed on 28 November 2020).
8. Witkowski, Ł.; Wiśniewski, J. *Ustawa o Elektromobilności i Paliwach Alternatywnych: Przewodnik infograficzny po wybranych zagadnieniach*. 2018. Available online: http://pspa.com.pl/assets/uploads/2018/11/RAPORT_PSPA_Przewodnik_po_ustawie_o_elektromobilnosci.pdf (accessed on 28 November 2020).
9. Ministry of Energy. *Electromobility Development Plan in Poland: Energy for the Future*; 2017. Available online: https://www.gov.pl/documents/33372/436746/DIT_PRE_EN.pdf (accessed on 28 November 2020).
10. Ziemba, P. Towards Strong Sustainability Management—A Generalized PROSA Method. *Sustainability* **2019**, *11*, 1555. [CrossRef]
11. Ziemba, P.; Wątróbski, J.; Zioło, M.; Karczmarczyk, A. Using the PROSA Method in Offshore Wind Farm Location Problems. *Energies* **2017**, *10*, 1755. [CrossRef]

12. Domingues, A.R.; Marques, P.; Garcia, R.; Freire, F.; Dias, L.C. Applying Multi-Criteria Decision Analysis to the Life-Cycle Assessment of vehicles. *J. Clean. Prod.* **2015**, *107*, 749–759. [CrossRef]
13. Sałabun, W.; Karczmarczyk, A. Using the COMET Method in the Sustainable City Transport Problem: An Empirical Study of the Electric Powered Cars. *Procedia Comput. Sci.* **2018**, *126*, 2248–2260. [CrossRef]
14. Ziemba, P. Inter-Criteria Dependencies-Based Decision Support in the Sustainable wind Energy Management. *Energies* **2019**, *12*, 749. [CrossRef]
15. Barfod, M.B.; Kaplan, S.; Frenzel, I.; Klauenberg, J. COPE-SMARTER—A decision support system for analysing the challenges, opportunities and policy initiatives: A case study of electric commercial vehicles market diffusion in Denmark. *Res. Transp. Econ.* **2016**, *55*, 3–11. [CrossRef]
16. Adhikari, M.; Ghimire, L.P.; Kim, Y.; Aryal, P.; Khadka, S.B. Identification and Analysis of Barriers against Electric Vehicle Use. *Sustainability* **2020**, *12*, 4850. [CrossRef]
17. Zhang, L.; Zhao, Z.; Chai, J.; Kan, Z. Risk Identification and Analysis for PPP Projects of Electric Vehicle Charging Infrastructure Based on 2-Tuple and the dematel Model. *World Electr. Veh. J.* **2019**, *10*, 4. [CrossRef]
18. Liu, J.; Wei, Q. Risk evaluation of electric vehicle charging infrastructure public-private partnership projects in China using fuzzy TOPSIS. *J. Clean. Prod.* **2018**, *189*, 211–222. [CrossRef]
19. Fazeli, R.; Davidsdottir, B.; Shafiei, E.; Stefansson, H.; Asgeirsson, E.I. Multi-criteria decision analysis of fiscal policies promoting the adoption of electric vehicles. *Energy Procedia* **2017**, *142*, 2511–2516. [CrossRef]
20. Erbaş, M.; Kabak, M.; Özceylan, E.; Çetinkaya, C. Optimal siting of electric vehicle charging stations: A GIS-based fuzzy Multi-Criteria Decision Analysis. *Energy* **2018**, *163*, 1017–1031. [CrossRef]
21. Ju, Y.; Ju, D.; Santibanez Gonzalez, E.D.R.; Giannakis, M.; Wang, A. Study of site selection of electric vehicle charging station based on extended GRP method under picture fuzzy environment. *Comput. Ind. Eng.* **2019**, *135*, 1271–1285. [CrossRef]
22. Xu, F.; Liu, J.; Lin, S.; Yuan, J. A VIKOR-based approach for assessing the service performance of electric vehicle sharing programs: A case study in Beijing. *J. Clean. Prod.* **2017**, *148*, 254–267. [CrossRef]
23. Wątróbski, J.; Małecki, K.; Kijewska, K.; Iwan, S.; Karczmarczyk, A.; Thompson, R.G. Multi-Criteria Analysis of Electric Vans for City Logistics. *Sustainability* **2017**, *9*, 1453. [CrossRef]
24. Wątróbski, J.; Jankowski, J.; Ziemba, P.; Karczmarczyk, A.; Zioło, M. Generalised framework for multi-criteria method selection. *Omega* **2019**, *86*, 107–124. [CrossRef]
25. Rowley, H.V.; Peters, G.M.; Lundie, S.; Moore, S.J. Aggregating sustainability indicators: Beyond the weighted sum. *J. Environ. Manag.* **2012**, *111*, 24–33. [CrossRef] [PubMed]
26. Moghaddam, N.B.; Nasiri, M.; Mousavi, S.M. An appropriate multiple criteria decision making method for solving electricity planning problems, addressing sustainability issue. *Int. J. Environ. Sci. Technol.* **2011**, *8*, 605–620. [CrossRef]
27. Chen, S.-J.; Hwang, C.-L. *Fuzzy Multiple Attribute Decision Making: Methods and Applications*; Lecture Notes in Economics and Mathematical Systems; Springer: Berlin/Heidelberg, Germany, 1992; ISBN 978-3-540-54998-7.
28. Havranek, T. Multi-criteria decision analysis for environmental remediation: Benefits, challenges, and recommended practices. *Remediat. J.* **2019**, *29*, 93–108. [CrossRef]
29. Guitouni, A.; Martel, J.-M. Tentative guidelines to help choosing an appropriate MCDA method. *Eur. J. Oper. Res.* **1998**, *109*, 501–521. [CrossRef]
30. Cinelli, M.; Coles, S.R.; Kirwan, K. Analysis of the potentials of multi criteria decision analysis methods to conduct sustainability assessment. *Ecol. Indic.* **2014**, *46*, 138–148. [CrossRef]
31. Brans, J.-P.; De Smet, Y. PROMETHEE Methods. In *Multiple Criteria Decision Analysis: State of the Art Surveys*; Greco, S., Ehrgott, M., Figueira, J.R., Eds.; International Series in Operations Research & Management Science; Springer: New York, NY, USA, 2016; pp. 187–219. ISBN 978-1-4939-3094-4.
32. Ziemba, P.; Jankowski, J.; Wątróbski, J. Dynamic Decision Support in the Internet Marketing Management. In *Transactions on Computational Collective Intelligence XXIX*; Lecture Notes in Computer Science; Springer: Berlin/Heidelberg, Germany, 2018; pp. 39–68. ISBN 978-3-319-90286-9.
33. Ishizaka, A.; Nemery, P. *Multi-criteria Decision Analysis: Methods and Software*, 1st ed.; Wiley: Chichester, UK, 2013; ISBN 978-1-119-97407-9.
34. Chen, J.; Wang, J.; Baležentis, T.; Zagurskaitė, F.; Streimikiene, D.; Makutėnienė, D. Multicriteria Approach towards the Sustainable Selection of a Teahouse Location with Sensitivity Analysis. *Sustainability* **2018**, *10*, 2926. [CrossRef]

35. Pinto, G.; Abdollahi, E.; Capozzoli, A.; Savoldi, L.; Lahdelma, R. Optimization and Multicriteria Evaluation of Carbon-neutral Technologies for District Heating. *Energies* **2019**, *12*, 1653. [CrossRef]
36. Tervonen, T.; Lahdelma, R. Implementing stochastic multicriteria acceptability analysis. *Eur. J. Oper. Res.* **2007**, *178*, 500–513. [CrossRef]
37. Wang, J.; Luo, P.; Hu, X.; Zhang, X. Combining an Extended SMAA-2 Method with Integer Linear Programming for Task Assignment of Multi-UCAV under Multiple Uncertainties. *Symmetry* **2018**, *10*, 587. [CrossRef]
38. Wiśniewski, J.; Kania, A. *Katalog Pojazdów Elektrycznych 2019/2020. Wydanie II*; Polish Alternative Fuels Association: Warszawa, Poland, 2019.
39. Kwinta, W. Pierwsze Samochody Elektryczne dla Krakowskich Instytucji Miejskich Już są. Available online: https://inzynieria.com/energetyka/wiadomosci/56970,pierwsze-samochody-elektryczne-dla-krakowskich-instytucji-miejskich-juz-sa (accessed on 28 November 2020).
40. ORPA. Centrum Obsługi Administracji Rządowej ogłosiło przetargi na zakup i wynajem samochodów elektrycznych. 2020. Available online: https://orpa.pl/centrum-obslugi-administracji-rzadowej-oglosilo-przetargi-na-zakup-i-wynajem-samochodow-elektrycznych-i-hybryd/ (accessed on 28 November 2020).
41. Rynek Infrastruktury Szczecin Stawia na Elektromobilność. Prezydent Też w Elektryku. Available online: https://www.rynekinfrastruktury.pl/wiadomosci/drogi/prezydent-szczecina-jezdzi-elektrykiem-miasto-stawia-na-elektromobilnosc-67022.html (accessed on 28 November 2020).
42. ZDM Kupimy 4 Samochody Elektryczne. Available online: https://zdm.waw.pl/aktualnosci/kupimy-4-samochody-elektryczne/ (accessed on 28 November 2020).
43. NCAP. The European New Car Assessment Programme | Euro NCAP. Available online: https://www.euroncap.com:443/en (accessed on 28 November 2020).
44. EV Database. Available online: ev-database.org (accessed on 28 November 2020).
45. Abastante, F.; Corrente, S.; Greco, S.; Ishizaka, A.; Lami, I.M. A new parsimonious AHP methodology: Assigning priorities to many objects by comparing pairwise few reference objects. *Expert Syst. Appl.* **2019**, *127*, 109–120. [CrossRef]
46. Salminen, P.; Hokkanen, J.; Lahdelma, R. Comparing multicriteria methods in the context of environmental problems. *Eur. J. Oper. Res.* **1998**, *104*, 485–496. [CrossRef]
47. Hyde, K.; Maier, H.R.; Colby, C. Incorporating uncertainty in the PROMETHEE MCDA method. *J. Multi-Criteria Decis. Anal.* **2003**, *12*, 245–259. [CrossRef]
48. Deshmukh, S.C. Preference Ranking Organization Method Of Enrichment Evaluation (Promethee). *Int. J. Eng. Sci. Invent.* **2013**, *2*, 28–34.
49. Roy, B. The outranking approach and the foundations of electre methods. *Theor. Decis.* **1991**, *31*, 49–73. [CrossRef]
50. Amponsah, S.K. Logistic preference function for preference ranking organization method for enrichment evaluation (PROMETHEE) decision analysis. *Afr. J. Math. Comput. Sci. Res.* **2012**, *5*, 112–119. [CrossRef]
51. Podvezko, V.; Podviezko, A. *Use and Choice of Preference Functions for Evaluation of Characteristics of Socio-Economical Processes*; Vilnius Gediminas Technical University: Vilnius, Lithuania, 2010; pp. 1066–1071.
52. Podvezko, V.; Podviezko, A. Dependence of multi-criteria evaluation result on choice of preference functions and their parameters. *Technol. Econ. Dev. Econ.* **2010**, *16*, 143–158. [CrossRef]
53. Ziemba, P. NEAT F-PROMETHEE—A new fuzzy multiple criteria decision making method based on the adjustment of mapping trapezoidal fuzzy numbers. *Expert Syst. Appl.* **2018**, *110*, 363–380. [CrossRef]
54. Ziemba, P.; Becker, J. Analysis of the Digital Divide Using Fuzzy Forecasting. *Symmetry* **2019**, *11*, 166. [CrossRef]
55. Ziemba, P.; Jankowski, J.; Wątróbski, J. Online Comparison System with Certain and Uncertain Criteria Based on Multi-criteria Decision Analysis Method. In Proceedings of the 9th International Conference on Computational Collective Intelligence, Nicosia, Cyprus, 27–29 September 2017; Nguyen, N.T., Papadopoulos, G.A., Jędrzejowicz, P., Trawiński, B., Vossen, G., Eds.; Springer International Publishing: Berlin/Heidelberg, Germany, 2017; pp. 579–589.
56. Ziemba, P.; Becker, A.; Becker, J. A Consensus Measure of Expert Judgment in the Fuzzy TOPSIS Method. *Symmetry* **2020**, *12*, 204. [CrossRef]

57. Elektrowóz.pl Nabór wniosków o dopłaty do aut elektrycznych "Zielony Samochód.". *SAMOCHODY ELEKTRYCZNE*. 2020. Available online: www.elektrowoz.pl (accessed on 28 November 2020).
58. Szymaczek, M. Ruszają dopłaty do aut elektrycznych, ale jest haczyk—w auto motor i sport. Available online: https://www.auto-motor-i-sport.pl/wydarzenia/Ruszaja-doplaty-do-aut-elektrycznych-ale-jest-haczyk,40672,1 (accessed on 28 November 2020).

Article

Substantiation of Loading Hub Location for Electric Cargo Bikes Servicing City Areas with Restricted Traffic†

Vitalii Naumov

Faculty of Civil Engineering, Cracow University of Technology, Warszawska 24, 31155 Krakow, Poland; vnaumov@pk.edu.pl; Tel.: +48-12-374-30-83

† The present work is the extended version of the paper "Choosing the localisation of loading points for the cargo bicycles system in the Krakow Old Town" presented at RelStat 2018 Conference, Riga, Lavia, 17–20 October 2018; and published in Lecture Notes in Networks and Systems.

Abstract: Electric cargo bicycles have become a popular mode of transport for last-mile goods deliveries under conditions of restricted traffic in urban areas. The indispensable elements of the cargo bike delivery systems are loading hubs: they serve as intermediate points between vans and bikes ensuring loading, storage, and e-vehicle charging operations. The choice of the loading hub location is one of the basic problems to be solved when designing city logistics systems that presume the use of electric bicycles. The paper proposes an approach to justifying the location of a loading hub based on computer simulations of the delivery process in the closed urban area under the condition of stochastic demand for transport services. The developed mathematical model considers consignees and loading hubs as vertices in the graph representing the transport network. A single request for transport services is described based on the set of numeric parameters, among which the most significant are the size of the consignment, its dimensions, and the time interval between the current and the previous requests for deliveries. The software implementation of the developed model in Python programming language was used to simulate the process of goods delivery by e-bikes for two cases—the synthetically generated rectangular network and the real-world case of the Old Town district in Krakow, Poland. The loading hub location was substantiated based on the simulation results from a set of alternative locations by using the minimum of the total transport work as the efficiency criterion. The obtained results differ from the loading hub locations chosen with the use of classical rectilinear and center-of-gravity methods to solve a simple facility location problem.

Keywords: cargo bicycles; loading hub; facility location problem; computer simulations; Python programing

Citation: Naumov, V. Substantiation of Loading Hub Location for Electric Cargo Bikes Servicing City Areas with Restricted Traffic. *Energies* **2021**, *14*, 839. https://doi.org/10.3390/en14040839

Academic Editor: Albert Y.S. Lam
Received: 5 January 2021
Accepted: 2 February 2021
Published: 5 February 2021

1. Introduction

The growth of e-commerce during the last decade is one of the main reasons for the exponentially increased demand for last-mile deliveries that are characterized by cargo diversity and low utilization of vehicles' capacity. Commercial vehicles delivering the packages directly to customers may significantly contribute to the increase of traffic in urban areas. Especially, this concerns the central city areas with historic buildings and low capacity of the roads: commercial vehicles are the main source of noise and air pollution in such areas, they cause congestions and reduce the public space. For these reasons, nowadays, the restriction of traffic in the selected districts is provided in many cities: the motorized vehicles may enter the area only during the specified time window (usually these are night or early-morning hours). Such restrictions, however, negatively influence the businesses located in the areas with restricted traffic by increasing the storage costs and causing the loss of profit due to the lack of goods at the given time moment. In such a situation, cargo distribution systems based on cargo bikes could be considered as the alternative to conventional last-mile deliveries by motorized vehicles.

Loading hubs are necessary elements of the urban distribution system based on cargo bicycles: loads are transported by vans to the storage point located at the bound of the area with restricted traffic and after are delivered to consignees. As examples of such hubs, the following real-world implementations may be listed: a trailer or a semi-trailer, a cargo container unit used as a mobile point, the dedicated transshipment bay, a parcel locker, or a storage space having the character of a permanent loading point. As the core function of the loading hubs, besides the transshipment operations and short-term storage, the e-bikes charging operations should be mentioned.

Both from practical and theoretical points of view, choosing the loading hub location constitutes a significant problem. Practically, such alternative locations should be found that satisfy the legislative restrictions and are characterized by availability for conventional transport and e-bikes. Then, theoretically, by applying a chosen methodology, such the hub location should be distinguished from the set of alternative variants that guarantees the most effective operation of the distribution system.

The current study presents the novel simulation-based methodology to choose the location of a loading hub for a cargo bicycle system given the set of alternative locations. The developed approach, unlike the existing simulation-based methods, considers stochasticity of demand for deliveries in the selected urban area as well as allows the implementation of the routing procedures as part of the technology of goods delivering by electric cargo bikes.

The paper has the following structure: The second part presents a brief review of recent publications related to the research problem; the proposed methodology to substantiate the loading hub location for a cargo bikes delivery system is described in the third part; the fourth part contains the introductive description of software implementing the developed model; the fifth section introduces case studies of choosing the loading point location for a synthetic rectangular network and the cargo bikes delivery system in the Old Town district of Krakow, Poland; the last part offers brief conclusions and directions of future research.

2. Literature Review

Many recent studies underline the advantages of cargo bicycles as the sustainable mean of transport under urban conditions [1–3]. The author of the paper [4] concludes that the use of cargo bikes contributes to the reduction of traffic congestion and the improvement of air quality. According to the publication [5], both operational and external transport costs can be reduced if the delivery scheme with transshipment hubs and cargo bikes is used for the distribution of e-commerce goods in Antwerp (Belgium). The authors of the research [6] point out that the use of electric cargo bikes in urban logistics activities in Porto (Portugal) reach up to 25% of reductions in external costs. Numeric results regarding the decrease of CO_2 emissions due to the use of bicycles for cargo deliveries also show the significant impact: According to [6], CO_2 emissions can be reduced by up to 73%, while the authors of the research [7] state that the use of electrically assisted cargo tricycles results in the CO2 emissions decrease by 54% per parcel delivered. Authors of the publication [6] also conclude that cargo bikes can replace up to 10% of the conventional vans without changing the overall network efficiency. The study [8] shows that the expected travel time for delivery distances up to 20 km is on average just 6 min longer if cargo bikes are used instead of traditional vans.

The transport of loads by bikes also has many limitations related to the delivery distance, the consignment size, the vehicles' speed, and the safety of the goods. However, these limitations are not subject to the case of deliveries under the urban conditions when electrically powered bicycles are used: Delivery distances are relatively short, the e-bike's capacity is comparable with the capacity of light commercial vehicles, the speed of conventional vans does not exceed the average e-bikes' speed due to road traffic restrictions, and the freight is protected from the weather conditions when the bikes with the covered bins are used.

The contemporary scientific literature describes the set of operational problems to be solved while designing the e-bike delivery system in an urban area: vehicle routing [9–11], synchronization of bikes with conventional vans [12], choice of vehicle type [10,13], vehicles scheduling [14,15], shaping the recharging strategy [15–17], etc. However, the most significant practical and theoretical issue when setting up the regional e-bike delivery system is the substantiation of the transshipment hub(s) parameters, including the type and location of the hub(s).

As it pointed out in the study [13], contemporary knowledge about planning urban transshipment points is very limited. Substantiation of the loading hub location refers to the branch of the operations research known as the facility location problem (FLP) [18]. The simplest version of FLP (the Weber problem) assumes the choice of a single facility location for the given set of customers' locations. For the multiple hub locations, the FLP becomes NP-hard and is usually solved by using different approximation algorithms that find the solution within a reasonable computational time. In real-world applications of FLP, the uncertainty of demand for deliveries, as well as the stochastic parameters of the servicing process and resource restrictions, should be considered [19–23]; it significantly complicates the process of searching for a reliable solution.

To solve FLP, various approaches are used by scholars and practitioners. The most popular mathematical apparatus for substantiation of the loading hubs location is the mixed-integer linear programming [9,14,15,21,24], although the fuzzy-logic-based models are used as well [17,25]. Another direction for finding the solution of FLP is the simulation-based approaches [5,8,19,26]: The delivery and servicing processes are simulated for a set of alternative locations and the best option is being substantiated based on the simulation results. The simulation-based approaches assume that the adequate model of the transport system is developed as the base for simulations. The model is used to run multiple simulations of the transport system to consider the overall influence of random parameters on the simulations' results and to evaluate statistically significant values of the efficiency criteria. The authors of the study [27] use the robust optimization approach to alleviate the impact of uncertainty on obtained solutions.

Different heuristics are used to obtain the solution for FLP. The authors of the paper [28], to place bicycle stations in a way that minimizes the distance between clients and the closest bike station in Malaga (Spain), have studied the comparative advantages of the following algorithms: genetic algorithm, iterated local search, particle swarm optimization, simulated annealing, and variable neighborhood search. Genetic algorithms (GA) are quite a popular heuristic for solving FLP: e.g., the optimal location and size of charging stations are substantiated based on the developed GA in the study [17], the location of a transshipment hub is being chosen based on the proposed GA in the paper [29] considering the vehicle routing problem in the multistage distribution system.

Although the contemporary literature proposes a variety of methods to substantiate the loading hub location, it should be mentioned that existing approaches are usually case-study-sensitive and allow solving the problem for the specific situation or the given initial data. Considering the presented literature review, the following research gaps should be filled by the proposed simulation-based approach:

- the demand for deliveries is presented by deterministic parameters: the stochastic nature of the demand for transport services should be considered, that may be achieved by implementing the demand simulation procedure within the model of a distribution system;
- the technological procedures are not considered in detail: the result of some servicing operations (such as vehicle routing or loads scheduling) depend on the demand parameters; conventional linear programming models cannot consider this; however, the corresponding procedures may be implemented within the simulation models.

3. Simulation-Based Methodology for the Substantiation of a Loading Hub Location

As the basis for estimating the alternative location of the loading hub, the simulation model of the transportation system is proposed to be used. Such model simulates demand for transport services, as well as the technological processes: handling and charging operations in loading hubs, delivery of consignments to consignees located in the serviced area, as well as loading and unloading operations at customers' locations and in a hub. To run the model, its proper program implementation should be developed. Based on the simulation results and considered efficiency criteria, the expediency of the loading hub locations is estimated.

The process of goods delivery by electric bikes is implemented within the logistic system of the considered urban area. As basic structural elements of this system, the following subsystems should be listed:

- the transport network of the selected urban area where the processes of goods delivery take place;
- demand for the supply of goods reflecting the customers' needs for deliveries;
- the subsystem serving the transport demand that includes means of transport (cargo bicycles) and loading hubs.

Thus, the mathematical model $\mathbf{M}_{DS}$ of the delivery system in a general form can be presented as the tuple of three above listed elements:

$$\mathbf{M}_{DS} = \langle \mathbf{\Omega},\ \mathbf{D},\ \mathbf{\Phi} \rangle, \tag{1}$$

where $\mathbf{\Omega}$ is the transport network; $\mathbf{D}$ is demand for transport services; $\mathbf{\Phi}$ is the servicing subsystem.

3.1. Transport Network Model

The most commonly used approach to modeling transport networks is the use of mathematical structures based on graph models. The transport network can be formalized as a pair of subsets—the nodes and the edges:

$$\mathbf{\Omega} = \langle \{\eta_i\},\ \{\lambda_j\} \rangle,\ \forall \eta_i \in \mathbf{N},\ \forall \lambda_j \in \mathbf{\Lambda}, \tag{2}$$

where η_i is the i-th node being the element of the network nodes set $\mathbf{N}$; λ_j is the j-th edge being the element of the network edges set $\mathbf{\Lambda}$.

In the proposed mathematical model, as the nodes of the network, the locations of potential customers of transport services are presented, while the edges (links) are used to reflect the relevant sections of the road network connecting the nodes. Locations of the loading hubs, from which the cargoes are delivered by bikes, can be also represented as the network vertices. Additionally, for the more detailed networks, the nodes can be used to model the road intersections.

The basic characteristics of a node are its geographical coordinates and lists of ingoing and outgoing edges:

$$\eta = \langle x,\ y,\ \mathbf{\lambda}_{in},\ \mathbf{\lambda}_{out} \rangle, \tag{3}$$

where x and y are the coordinates characterizing the node location (latitude and longitude may be used as such coordinates); $\mathbf{\lambda}_{in}$ and $\mathbf{\lambda}_{out}$ are the sets of ingoing and outgoing edges for the given node η : $\mathbf{\lambda}_{in} \subset \mathbf{\Lambda}$, $\mathbf{\lambda}_{out} \subset \mathbf{\Lambda}$.

The obligatory set of parameters describing the graph edge includes its weight and end vertices:

$$\lambda = \langle w,\ \eta_{out},\ \eta_{in} \rangle, \tag{4}$$

where w is the edge weight (e.g., the length of the corresponding road section, the travel time, the transport costs, etc.); η_{out} and η_{in} are the beginning and the ending nodes of the edge: $\eta_{out} \in \mathbf{N}$, $\eta_{in} \in \mathbf{N}$.

Note that in a more advanced version of the network model, the set of node and edge parameters could be extended depending on the problem being solved.

3.2. Model of Demand for Goods Delivery

The core unit shaping transport demand is a request for the delivery of goods understood as the customer's need for services, supported by his purchasing ability, and presented on the market to be satisfied. A set of requests for the services of a given company forms the demand for the services of a transport enterprise. Each request could be quantified based on a set of numerical parameters.

In the general form, the demand model for goods delivery can be presented as an ordered set of requests:

$$\mathbf{D} = \{\rho_1, \rho_2, \ldots, \rho_N\}, \tag{5}$$

where ρ_i is the i-th request in the flow: $\rho_i \prec \rho_{i+1}$ if $t_i \leq t_{i+1}$, t_i is the moment of appearance of the i-th request; N is the number of requests in a flow.

A single request for delivery is characterized by the following parameters:

$$\rho = \langle \eta_o, \eta_d, \zeta, \omega, \langle \theta_l, \theta_w, \theta_h \rangle \rangle, \tag{6}$$

where η_o and η_d are the transport network nodes defining the location of a sender and a recipient: $\eta_o \in \mathbf{N}$, $\eta_d \in \mathbf{N}$; ζ is the time interval between the moments of appearance of the given and previous requests [min]; ω is the consignment weight [kg]; θ_l, θ_w, and θ_h are dimensions of the load unit—its length, width, and height [cm].

In the Equation (6), the time interval ζ characterizes the intensity of the demand for deliveries for the given consignor: the more the value of the interval, the less intense the demand for transport services.

For a flow of requests with a finite number of elements, the subsetting of requests by sender and recipient can be presented in the form of a travel matrix $\mathbf{\Delta}$. An element δ_{ij} of such the matrix reflects the number of requests for which the sender is located in the node η_i (origin node), while the recipient—in the node η_j (destination node).

For a single request, its numerical parameters are deterministic, but for the requests flow, these parameters are random, and can be described by stochastic variables. It follows that the demand model for cargo transport can be presented as a set of random variables characterizing the numerical parameters of requests for cargo deliveries supported by the origin-destination matrix defining spatial distribution of the requests:

$$\mathbf{D} = \left\langle \mathbf{\Delta}, \tilde{\zeta}, \tilde{\omega}, \left\langle \tilde{\theta}_l, \tilde{\theta}_w, \tilde{\theta}_h \right\rangle \right\rangle, \tag{7}$$

where $\tilde{x}$ is a random variable characterizing some numeric parameter x of demand for cargo deliveries.

Given the set of ordered requests for the cargo deliveries characterize the demand, the task of demand modeling can be presented as the task of generating numerical parameters of the requests in a flow. Implementation of the demand model as a flow of N elements is a set of requests in the form (5), where numerical parameters are values from samples, generated for the respective random variables, and the locations of senders and recipients are defined by the matrix $\mathbf{\Delta}$.

3.3. Model of the Servicing Subsystem

Basic components of the servicing subsystem are the fleet of vehicles (cargo bicycles) and a set of loading hubs:

$$\mathbf{\Phi} = \langle \{b_i\}, \{p_j\} \rangle, \forall b_i \in \mathbf{B}, \forall p_j \in \mathbf{P}, \tag{8}$$

where b_i is the i-th cargo bicycle being the element of the vehicles fleet $\mathbf{B}$; p_j is the j th loading hub being the element of all the loading hubs' set $\mathbf{P}$ (in case, when the single hub location problem is solved, the set $\mathbf{P}$ contains just one element).

The basic parameters of cargo bikes as elements of the servicing system are their load capacity, dimensions of the cargo space, and technical speed:

$$b = \langle q_b, v_b, \langle l_b, w_b, h_b \rangle \rangle, \tag{9}$$

where q_b is the load capacity of a cargo bicycle [kg]; v_b is the average technical speed of a cargo bicycle [km/h] (the technical speed may be also considered as a random variable); l_b, w_b, and h_b are dimensions of the bicycle cargo space—its length, width, and height [cm].

Note that the battery capacity may be considered as the key parameter for electric bikes, if it is used as the restriction in the problem being solved or if it is needed for the estimation of the resulting efficiency criteria. The dimensions of the loading space are considered in the model as the additional restriction to the load capacity (these parameters may be used if the solution of the knapsack problem is being considered in the procedure implementing the loading operations).

Loading hubs within the frame of the proposed model are short-term (temporary) storage points to which cargo is transported by other means of transport (usually—by delivery vans) to deliver them to the final recipient. The basic characteristics of loading hubs are their location and capacity:

$$p = \langle \eta_p, q_p \rangle, \tag{10}$$

where η_p is the node of the transport network representing the location of the loading hub, $\eta_p \in \mathbf{N}$; q_p is the loading hub capacity (possible maximum amount of cargo that can be stored at the hub) [t or m^3].

3.4. *Shaping the Routes for Delivery of Cargoes by Electric Bikes*

In most cases, the deliveries of goods by bicycles are carried out under conditions of combined consignments: By aiming the minimization of the total distance, the delivery routes are being formed for the requests that are received during the considered time window (in such situations, the total weight of the combined consignment should not exceed the vehicle's capacity). Assuming that servicing companies behave rationally and shape rational (or optimal) delivery routes while servicing the clients, the simulation model should approximate the real-world behavior of transport operators, and thus—it should contain the implementation of routing procedures.

As input data in the routing algorithms, the matrix of the shortest distances between the vertices of the transport network is used. To estimate such matrix, any known method of searching for the shortest paths should be used (e.g., Dijkstra algorithm or Floyd–Warshall algorithm [30]). For the generated flow of requests, the task of shaping the delivery routes is solved as the traveling salesman problem (TSP), where the location of the load sender (the parameter η_o) is defined as the location of a loading hub, and locations of recipients (the parameter η_d)—as the locations of the corresponding nodes in the transport network model. Known heuristic methods can be used for solving the TSP (e.g., Clarke–Wright algorithm, simulated annealing method, methods based on genetic algorithms, ant colony optimization methods, etc. [31]). The result of the routing procedure is a set of delivery routes, being the ordered sets of vertices of the graph representing the network model. The first and last element in the set of nodes shaping the route are the vertices defining the location of a loading hub.

The basic characteristics of delivery routes, calculated based on given parameters of the mathematical model, are the route length (total distance covered), the total weight of transported cargo, and the total transport work. These technological characteristics may be considered as key indicators to estimate the efficiency of the routing procedures, as well as the efficiency of the delivery system for alternative facility locations.

Note that time windows may be added to the model as additional restrictions: for each consignor, the time window when a parcel should be delivered may be considered. More advanced optimization techniques should be used within the routing procedure in this case, as far as the solution for the capacitated vehicle routing problem with time windows should be found.

3.5. The Objective Function for the Hub Location Problem

Special scientific literature proposes various efficiency criteria for solving FLP. The most widely used criterion is total costs: In the research [5], the e-commerce distribution system based on cargo bikes and delivery points is assessed with total operational and external costs, the authors of the paper [9] propose to select cost-minimizing locations of parcel lockers for goods deliveries by cargo bikes, the study [15] uses total costs as the objective function to substantiate the location of recharging stations for electrical vehicles. However, some recently developed approaches define facility locations considering environmental pollution together with operational costs: The model developed in [10] uses the total costs and the environmental impacts to assess alternative configurations of urban consolidation centers, while the study [27] considers the combination of total costs and parameters of environmental pollution to select locations of refueling stations.

There are also other efficiency criteria mentioned in recent studies and used to substantiate the facility location. Authors of the publication [20] use total system travel time and total system net energy consumption for optimal positioning of dynamic wireless charging infrastructure for battery electric vehicles, while the minimization of the power loss in a distribution network is considered as the objective function in the study [17] when solving the problem of the charging stations location. The paper [24] proposes solving the multimodal capacitated hub location problem based on the profitability of alternative locations. The authors of the study [7] use the total distance traveled and the CO2 emissions per parcel delivered to substantiate the efficiency of an urban micro-consolidation center located in the delivery area where the deliveries are performed by electrically-assisted cargo tricycles and electric vans.

For solving the hub location problem based on the simulations of the delivery process, the minimization of total transport work can be used as the core objective function reflecting the technological operations performed within the logistic system of the considered urban area:

$$W_{tkm}(\eta_H) = \sum_{i=1}^{N_R} \sum_{j=1}^{N_{S(i)}} q_{ij} \times d_{ij} \rightarrow \min, \quad (11)$$

where η_H is the node of the transport network $\mathbf{\Omega}$ that represents the loading hub location, $\eta_H \in \mathbf{N_H}$, $\mathbf{N_H}$ is the set of possible loading hub locations, $\mathbf{N_H} \subset \mathbf{N}$; N_R is the number of delivery routes formed to service the transport demand in the considered city area; $N_{S(i)}$ is the number of segments at the i-th route, $N_{S(i)} \geq 2$; q_{ij} is the total weight carried by a vehicle at the j-th segment of the i-th route [tons]; d_{ij} is the distance covered by a vehicle at the j-th segment of the i-th route [km].

Note that $\sum_{j=1}^{N_{S(i)}} q_{ij} \times d_{ij}$ represents the total transport work performed by a vehicle at the i-th delivery route.

The proposed objective function may be used for solving the multiple hub location problem as well. In such a case, the subsets of locations representing alternative solutions should be formed prior (if the number of available alternative locations n is greater than the number k of hubs to be located, then the number of subsets is equal to the number of possible combinations $\binom{n}{k}$), and the total transport work should be estimated for each of the combinations.

Criterion (11) is the resulting technological parameter that may be used as the base for the calculation of other indicators. The performed ton-kilometers estimated at the level of delivery routes can serve for evaluation of total costs, environmental pollution, and energy consumption for a heterogeneous fleet of vehicles (total transport work may also be used to estimate the values of these indicators for homogenous fleet).

4. Software Implementation

To implement simulation models of the systems of goods transport by cargo bikes, a library of core classes was created. The library has been developed by using the Python programming language (Version 3.9, Manufacturer–Python Software Foundation), and it is available in an open-source repository [32]. The structure of the library is presented in Figure 1.

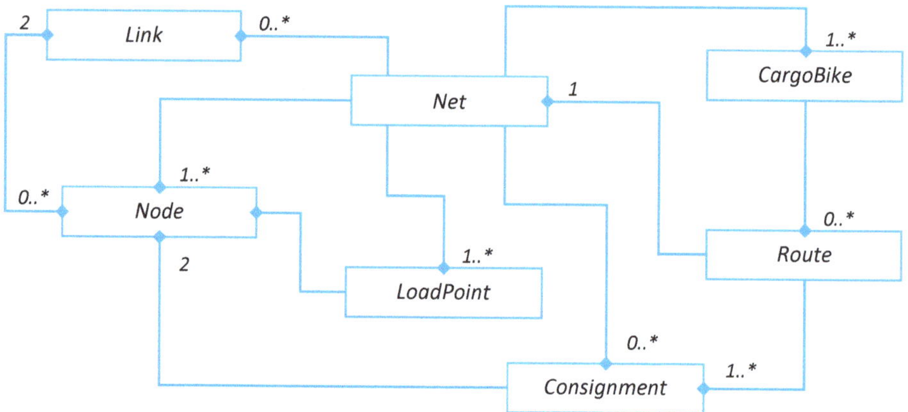

Figure 1. UML-diagram of the library for modeling of goods delivery by cargo bicycles.

The developed library contains the following core classes that are used to create a simulation model of the cargo bicycle transport system:

- *Net* class is used to implement the mathematical model of the transport network described in the previous section; this class contains lists of *Node* and *Link* elements defining the network configuration and the list of *Consignment* objects implementing the model of demand for transport services; the servicing system is implemented in the *Net* class as the lists of *LoadPoint* and *CargoBike* elements;
- *Node* and *Link* classes allow implementing the network vertices and edges as the models described by Equations (3) and (4);
- *Consignment* class is used to model requests for the delivery of goods; the sender and the consignee locations are defined within the given class as references to the appropriate *Node* objects of the transport network model;
- *Route* class allows implementing the program model of the delivery route; as far as the delivery route model can be defined only within a specific transport network, this class contains an object of the *Net* type being a reference to the network program model; the sequence of the route is defined as a list (ordered collection) of elements of the *Consignment* type; numeric parameters of the route (transport work, route length, total consignment weight) are calculated by methods implemented as properties of this class;
- *CargoBike* class is the program model of a cargo bicycle as the mean of transport used in the process of handling requests for the goods delivery;
- *LoadPoint* class is dedicated to developing program models of loading points; the point location is defined by the reference to the object of the *Node* type that represents the respective vertex of the transport network.

The *Net* class contains methods for generating demand as a flow of requests for the given random variables defining parameters of requests (consignment weight, the time interval between requests), the methods for calculating the shortest distance matrix based on the Floyd–Warshall and Dijkstra algorithm, as well as the methods for shaping delivery routes that implement the Clarke–Wright algorithm, the simulated annealing method, and the GA-based heuristic.

5. Case Studies and Discussion

The use of the proposed methodology of substantiating the loading hub location is presented for a synthetic transport network with two alternative locations and the real-world case study of the Old Town district in Krakow (Poland) with a set of feasible loading hub locations. The considered case studies represent examples of solving the FLP problem for a single facility.

5.1. Synthetic Network Example

Let us consider the rectangular (square) transport network, where the nodes define the locations of clients. To generate such the network, the dedicated procedure is implemented within the class *Net* [32]: the method takes as arguments the size of the network (the side of the square in nodes) and the random variable representing the length of the network edges. The Figure 2 shows the configuration of the square network with the square side equal to 7 nodes.

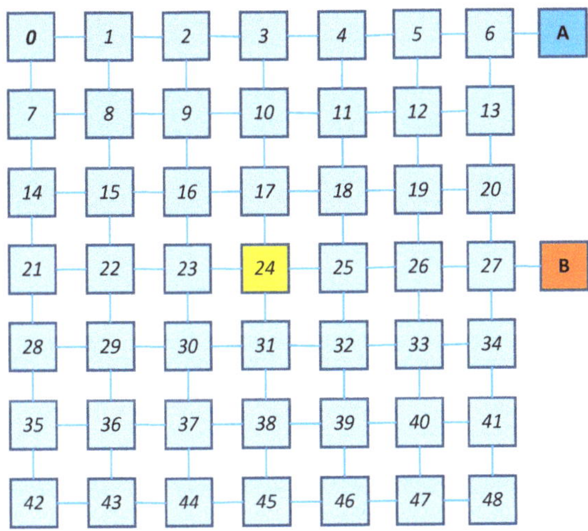

Figure 2. Synthetic network and alternative locations of loading hubs.

As alternative locations of the loading hub, the nodes connected to the corner node (location A) and the mid-side node (location B) are considered. The hub nodes are connected to the network with the edge that has the weight generated based on the same random variable that was used for the network generation.

To choose the loading hub locations out of two alternatives, the following simulation experiment was conducted:

- for each alternative location, 100 runs of the simulation model representing the delivery system were made,
- the distance between adjacent network vertices was generated as the random variable uniformly distributed between 50 and 200 m,
- the weight of shipments was generated as the normally distributed random variable with the average value of 30 kg and standard deviation of 5 kg,
- the probability of the request appearance (the probability that a client located in the given node has a request for goods delivery) is set equal to 0.5,
- total transport work at each launch of the simulation model was measured as the sum of ton-kilometers for all delivery routes obtained based on the Clarke–Wright heuristic (capacity of vehicles was set to 150 kg).

The results of the computer simulations are shown in Figure 3 as the distributions of random variables representing the total transport work conducted for each of the alternative variants of a loading hub location. The normal distribution of the random variables for both locations is confirmed with the Kolmogorov–Smirnov test for the probability of confidence equal to 0.95.

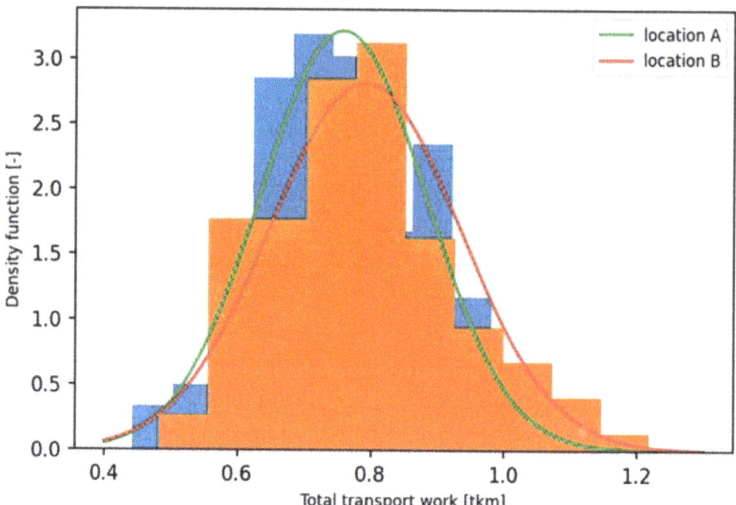

Figure 3. Distributions of the total transport work for the considered alternative hub locations in the synthetic network.

The obtained results of simulations show that the average total transport work for location A (0.757 tkm) is lower than the average value of this indicator for location B (0.789 tkm). On this basis, it can be concluded that the location of the loading hub in the corner of the rectangular network allows reducing the total transport work compared to the option of locating the hub in the middle of the rectangle side. However, it should be noted that such configurations of the network sections length and the parameters of demand are possible when the corner location of the transshipment point is characterized by a greater value of the objective function (in the conducted experiment, the 46 largest values of transport work for location A are greater than the corresponding lowest values of the objective function for location B).

A sufficient number of observations was evaluated for each of the experiment series based on the normal distribution of the objective function. For the significance level equal to 0.05, the sufficient number of the model runs is 29 for the series with location A and 35 for the series where a loading hub is located in node B. As the number of completed runs of the simulation model is bigger than the sufficient number of observations, it can be stated that the obtained results are statistically significant for the considered level of significance (the corresponding probability of confidence equal to 0.95).

The solution for the single-facility location problem obtained based on the rectilinear location model [18] for the considered example is the node located in the center of the network (that is node 24 highlighted at the net presented in Figure 2). Accordingly, when choosing between locations A and B, the node located closer to the center of the network would be a more preferable alternative (evidently, that is node B). As we can see, the solution obtained based on the proposed methodology differs from the conventional approach. This result may be explained by the effect of the routing procedures considered in the simulation model.

5.2. Kraków Old Town Case

The Old Town district of Krakow is the area with restricted traffic: only 3 streets of this district are not closed for motorized vehicles. At the same time, Krakow Old Town is the main touristic attraction of the region, where hundreds of restaurants, shops, and other touristic objects are located (locations of 730 commercial enterprises out of 954 objects registered in Google Maps are shown in Figure 4).

Figure 4. Locations of potential clients in the Old Town district of Krakow.

As far as the district is closed for heavy vehicles, the supply of the objects located in the district is allowed in the morning from 8 p.m. to 9.30 a.m. under the condition that only light good vehicles with the carrying capacity not exceeding 1.5 tons deliver the goods. Such restrictions lead to the increase of logistics expenses due to additional storage costs and the lack of the ability to implement on-time deliveries. The use of cargo bicycles is a solution that allows reducing logistics costs in this situation while preserving the tourist appeal of the district and low level of environmental pollution.

The location of a loading hub for the bikes delivery system in the Old Town of Krakow should be chosen from the set of alternative variants that satisfy the conditions of transport accessibility (both for conventional vehicles and cargo bicycles) and consider the architectural features of historic buildings in the district. The set of such alternative locations is presented in Table 1.

Table 1. The alternative locations of a loading hub in the Old Town district of Krakow.

Hub Location	Longitude	Latitude	Address of Location
A	50.05435	19.93880	St. Giles str.
B	50.06208	19.93341	St. Anna str.
C	50.06036	19.94142	Holy Cross str.
D	50.06416	19.93542	St. Stephen sq.
E	50.06364	19.94214	Holy Spirit sq.

Using the developed software, the transport network model that considers roads available for bicycles was prepared for the district of the Krakow Old Town. Initial data for the network model (locations of the objects and intersections, the type of the objects) were obtained by the means of Google Maps API. Enterprises and non-commercial objects located in the Krakow Old Town (including hotels, shops, restaurants, cafés, museums, and theaters) and the set of alternative locations of a loading hub were considered as the nodes of the graph representing the network model. The graph of the obtained network model is presented in Figure 5.

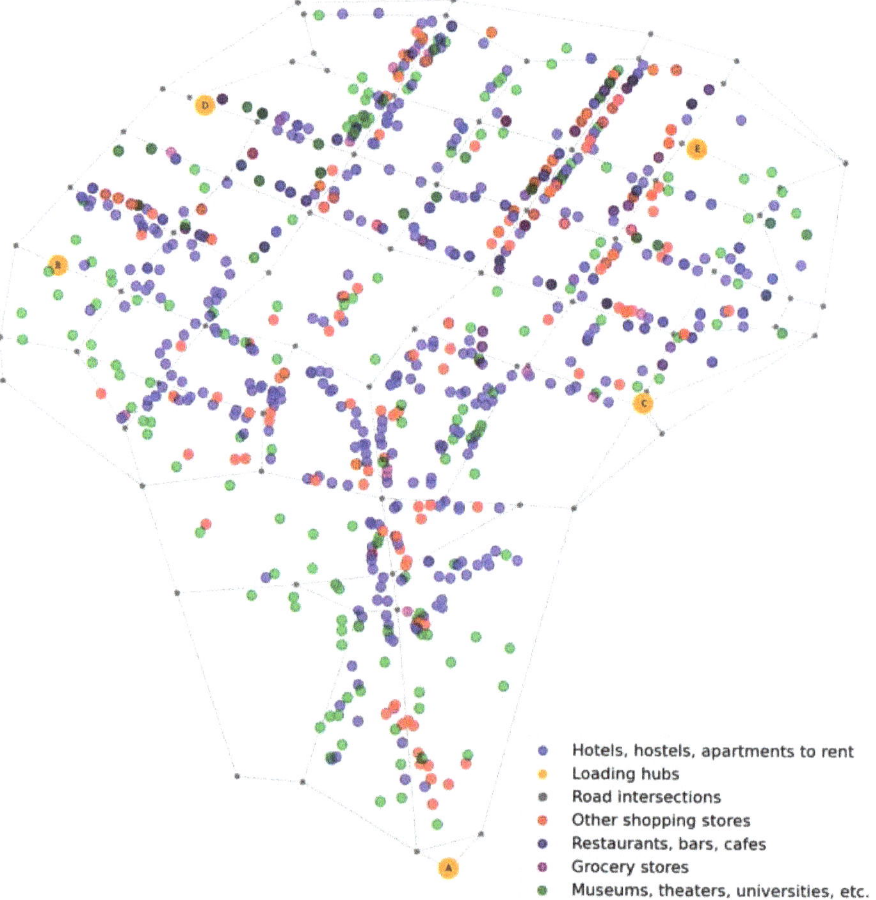

Figure 5. Locations of clients, intersections, and loading hubs in the Krakow Old Town.

To choose the best location of the loading hub, the Monte-Carlo simulations were performed based on the developed model. For each element of the considered set of possible hub locations, 30 runs of the following procedure were conducted:

- requests for goods delivery were generated for the set of the clients,
- the delivery routes were formed for the set of generated requests by using the Clarke–Wright algorithm,
- the total transport work was calculated for the obtained delivery routes.

For the demand simulation procedure, the probability of the request appearance was taken equal to 0.1 for each client and the random variable of the consignment weight was generated as the normally distributed variable with the average value equal to 30 kg and standard deviation equal to 5 kg.

As the result, the samples characterizing random variables of the total transport work were obtained for the alternative loading hub locations (distributions of the total ton-kilometers performed at the delivery routes are shown in Figure 6).

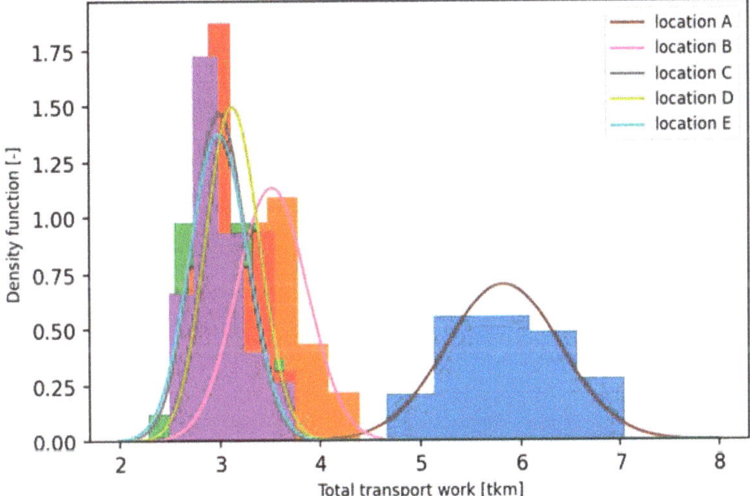

Figure 6. Distribution of the total transport work for the considered loading hub locations.

More detailed characteristics of the random variables representing the total transport work for the considered locations are presented in Table 2. To confirm that the obtained samples are big enough to make statistically significant conclusions, the sufficient number of observations was calculated for the level of significance equal to 0.05, assuming that the random variables representing the total transport work have a normal distribution. As it follows from the data presented in Table 2 (the number of performed observations is bigger than the corresponding sufficient number), the obtained results should be considered as statistically significant with the confidence probability equal to 0.95.

Based on the obtained results of computer simulations, locations C and E are the best options in the Krakow Old Town case, as they are characterized by the smallest mean value of the total transport work.

Table 2. Results of computer simulations for the considered loading hub locations.

Alternative Hub Location	Mean Value of Total Transport Work [tkm]	Variance of Total Transport Work [tkm^2]	Sufficient Number of Observations [obs.]
A	5.828	0.335	11
B	3.521	0.128	12
C	2.999	0.076	10
D	3.130	0.073	9
E	2.992	0.087	11

According to the center-of-gravity approach [18], for the set of potential clients located in the Old Town district of Krakow, the center of gravity is the point with coordinates (50.06178, 19.93846). Thus, based on the set of considered alternative locations, the solution of the single-facility location problem is the point closest to the center of gravity. The distances obtained from Google Maps (biking mode) and Euclidean distances between the gravity center and alternative hub locations are shown in Table 3.

Table 3. Distances from the gravity center to alternative hub locations.

Alternative Hub Location	Euclidean Distance [m]	Distance According to Google Maps [m]
A	744	850
B	506	500
C	328	300
D	386	450
E	412	400

As follows from the data presented in Table 3, both Euclidean and Google Maps distances to the gravity center are the smallest for location C, while location E is not the best alternative according to the conventional approach. Note that location A is evaluated as the worse possible option by the proposed methodology, as well as by the center-of-gravity approach.

The solution obtained by using the proposed simulation-based approach is superior to the one obtained based on the center-of-gravity approach: by choosing location E, we guarantee the minimum of the average total transport work (with 95% level of the probability of confidence), while by choosing location C, the minimum of the averaged values of the objective function cannot be guaranteed.

6. Conclusions

The proposed approach to the modeling of goods deliveries by cargo bicycles allows considering the stochastic nature of the demand for transport services and the used technology of the delivery process. The numerical parameters described in the mathematical model are the core characteristics, but the proposed model can be extended to be adapted for solving other optimization problems. Unlike the existing approaches, the proposed method allows taking into account the use of rational delivery routes, which increases the adequacy of the obtained results and the validity of the choice for the loading hub location problem. Since the main tool for choosing the loading hub location is the logistic system simulation, the proposed approach can be expanded by including other technological procedures in the program model in addition to routing.

By using the developed model and its software implementation, the problem of choosing the location for a loading point in the cargo bikes delivery system was solved for the Krakow Old Town district. It should be noted that the obtained results are preliminary: The detailed studies of demand and its simulation in the frame of the proposed model are needed for the real-world implementation of the system of goods delivery by cargo bicycles.

The use of the proposed approach in practice has the following implications:

- the chosen location of the transport hub will guarantee the minimum of total ton-kilometers performed by all e-bikes servicing the clients in the considered urban area; supposedly, the minimum of transport expenses will be achieved as well (for the homogenous fleet of vehicles);
- the stochastic nature of the demand for consignments delivery in the area will not affect the efficiency of the distribution system for chosen alternative location;
- the use of combined consignments will not deteriorate the resulting total transport work, as the rational behavior of the carriers regarding the used delivery routes is considered in the simulation model.

However, the use of the proposed methodology may have certain limitations placed upon its practical implementation:

- the parameters of demands should be studied prior for each group of potential clients located in the serviced area: for this, the surveys should be carried out aiming the estimation of the demand characteristics as stochastic variables; additionally, direct visual observations can be made for the selected clients to estimate the demand intensity (random variable of the time interval between requests in a flow);
- the method (or methods) of forming the delivery routes used by the carriers should be implemented within the simulation model: this limitation may be overcome directly if the carriers use a decision-support system that optimizes the delivery routes; otherwise, the behavior of the carriers should be studied to define the used routing strategies.

As the directions of further research on the developed methodology, the following topics should be mentioned:

- the development of the advanced procedures of demand simulations;
- the expansion of the model with algorithms of such technological operations as vehicle scheduling and stacking of load units in a bicycle cart;
- the study of the influence of different heuristics for solving the TSP problem on the simulation results for various parameters of demand for transport services;
- the consideration of additional restrictions related to the servicing subsystem, such as the number of available drivers and the drivers' schedule.

Funding: This research was funded by European Union's Horizon 2020 research and innovation program, grant agreement No. 769086 (CityChangerCargoBike project).

Institutional Review Board Statement: Not applicable.

Informed Consent Statement: Not applicable.

Data Availability Statement: The numeric results of the described simulations can be found at https://github.com/naumovvs/cargo-bikes-system.

Acknowledgments: The description of the developed model and the implemented software is reprinted by permission from Springer Nature Customers Service Center GmbH: Springer Lecture Notes in Networks and System, Choosing the localisation of loading points for the cargo bicycles system in the Krakow Old Town, Naumov V., Starczewski J. © 2019.

Conflicts of Interest: The author declares no conflict of interest.

References

1. Tipagornwong, C.; Figliozzi, M. Analysis of competitiveness of freight tricycle delivery services in urban areas. *Transp. Res. Rec.* **2014**, *2410*, 76–84. [CrossRef]
2. Schliwa, G.; Armitage, R.; Aziz, S.; Evans, J.; Rhoades, J. Sustainable city logistics–Making cargo cycles viable for urban freight transport. *Res. Transp. Bus. Manag.* **2015**, *15*, 50–57. [CrossRef]
3. Rudolph, C.; Gruber, J. Cargo cycles in commercial transport: Potentials, constraints, and recommendations. *Res. Transp. Bus. Manag.* **2017**, *24*, 26–36. [CrossRef]
4. Nürnberg, M. Analysis of using cargo bikes in urban logistics on the example of Stargard. *Transp. Res. Procedia* **2019**, *39*, 360–369. [CrossRef]

5. Arnold, F.; Cardenas, I.; Sörensen, K.; Dewulf, W. Simulation of B2C e-commerce distribution in Antwerp using cargo bikes and delivery points. *Eur. Transp. Res. Rev.* **2018**, *10*, 2. [CrossRef]
6. Melo, S.; Baptista, P. Evaluating the impacts of using cargo cycles on urban logistics: Integrating traffic, environmental and operational boundaries. *Eur. Transp. Res. Rev.* **2017**, *9*, 30. [CrossRef]
7. Browne, M.; Allen, J.; Leonardi, J. Evaluating the use of an urban consolidation centre and electric vehicles in central London. *IATSS Res.* **2011**, *35*, 1–6. [CrossRef]
8. Gruber, J.; Narayanan, S. Travel Time Differences between Cargo Cycles and Cars in Commercial Transport Operations. *Transp. Res. Rec.* **2019**, *2673*, 623–637. [CrossRef]
9. Enthoven, D.L.J.U.; Jargalsaikhan, B.; Roodbergen, K.J.; uit het Broek, M.A.J.; Schrotenboer, A.H. The two-echelon vehicle routing problem with covering options: City logistics with cargo bikes and parcel lockers. *Comput. Oper. Res.* **2020**, *118*, 104919. [CrossRef]
10. Simoni, M.D.; Bujanovic, P.; Boyles, S.D.; Kutanoglu, E. Urban consolidation solutions for parcel delivery considering location, fleet and route choice. *Case Stud. Transp. Policy* **2018**, *6*, 112–124. [CrossRef]
11. Naumov, V.; Vasiutina, H.; Starczewski, J. Web planning tool for deliveries by cargo bicycles in Kraków Old Town. *Adv. Intell. Syst. Comput.* **2020**, *1083*, 90–98. [CrossRef]
12. Anderluh, A.; Hemmelmayr, V.C.; Nolz, P.C. Synchronizing vans and cargo bikes in a city distribution network. *Cent. Eur. J. Oper. Res.* **2017**, *25*, 345–376. [CrossRef]
13. Assmann, T.; Lang, S.; Müller, F.; Schenk, M. Impact assessment model for the implementation of cargo bike transshipment points in urban districts. *Sustainability* **2020**, *12*, 4082. [CrossRef]
14. Alwesabi, Y.; Wang, Y.; Avalos, R.; Liu, Z. Electric bus scheduling under single depot dynamic wireless charging infrastructure planning. *Energy* **2020**, *213*, 118855. [CrossRef]
15. Wang, I.-L.; Wang, Y.; Lin, P.-C. Optimal recharging strategies for electric vehicle fleets with duration constraints. *Transp. Res. Part. C Emerg. Technol.* **2016**, *69*, 242–254. [CrossRef]
16. Xiao, D.; An, S.; Cai, H.; Wang, J.; Cai, H. An optimization model for electric vehicle charging infrastructure planning considering queuing behavior with finite queue length. *J. Energy Storage* **2020**, *29*, 101317. [CrossRef]
17. Liu, G.; Kang, L.; Luan, Z.; Qiu, J.; Zheng, F. Charging station and power network planning for integrated electric vehicles (EVs). *Energies* **2019**, *12*, 2595. [CrossRef]
18. Kasilingam, R.G. *Logistics and Transportation*; Springer Science: Berlin/Heidelberg, Germany, 1998; pp. 99–134.
19. Kchaou-Boujelben, M.; Gicquel, C. Locating electric vehicle charging stations under uncertain battery energy status and power consumption. *Comput. Ind. Eng.* **2020**, *149*, 106752. [CrossRef]
20. Ngo, H.; Kumar, A.; Mishra, S. Optimal positioning of dynamic wireless charging infrastructure in a road network for battery electric vehicles. *Transp. Res. Part. D Transp. Environ.* **2020**, *85*, 102385. [CrossRef]
21. Gao, H.; Liu, K.; Peng, X.; Li, C. Optimal location of fast charging stations for mixed traffic of electric vehicles and gasoline vehicles subject to elastic demands. *Energies* **2020**, *13*, 1964. [CrossRef]
22. Banerjee, S.; Kabir, M.M.; Khadem, N.K.; Chavis, C. Optimal locations for bikeshare stations: A new GIS based spatial approach. *Transp. Res. Interdiscip. Perspect.* **2020**, *4*, 100101. [CrossRef]
23. Efthymiou, D.; Antoniou, C.; Tyrinopoulos, Y. Spatially aware model for optimal site selection. *Transp. Res. Rec.* **2012**, *2276*, 146–155. [CrossRef]
24. Osorio-Mora, A.; Núñez-Cerda, F.; Gatica, G.; Linfati, R. Multimodal Capacitated Hub Location Problems with Multi-Commodities: An Application in Freight Transport. *J. Adv. Transp.* **2020**, *2020*, 2431763. [CrossRef]
25. Rao, C.; Goh, M.; Zhao, Y.; Zheng, J. Location selection of city logistics centers under sustainability. *Transp. Res. Part. D Transp. Environ.* **2015**, *36*, 29–44. [CrossRef]
26. Naumov, V.; Starczewski, J. Choosing the localisation of loading points for the cargo bicycles system in the Krakow Old Town. *Lect. Notes Netw. Syst.* **2019**, *68*, 353–362. [CrossRef]
27. Tafakkori, K.; Bozorgi-Amiri, A.; Yousefi-Babadi, A. Sustainable generalized refueling station location problem under uncertainty. *Sustain. Cities Soc.* **2020**, *63*, 102497. [CrossRef]
28. Cintrano, C.; Chicano, F.; Alba, E. Using metaheuristics for the location of bicycle stations. *Expert Syst. Appl.* **2020**, *161*, 113684. [CrossRef]
29. Szczepański, E.; Jacyna-Gołda, I.; Murawski, J. Genetic algorithms-based approach for transhipment hub location in urban areas. *Arch. Transp.* **2014**, *31*, 73–83. [CrossRef]
30. Simchi-Levi, D.; Chen, X.; Bramel, J. *The Logic of Logistics: Theory, Algorithms, and Applications for Logistics Management*; Springer: New York, NY, USA, 2014.
31. Labadie, N.; Prins, C.; Prodhon, C. *Metaheuristics for Vehicle Routing Problems*; John Wiley & Sons Inc.: London, UK, 2016.
32. Simulations of Goods Deliveries by Cargo Bicycles. Available online: https://github.com/naumovvs/cargo-bikes-system (accessed on 30 December 2020).

MDPI
St. Alban-Anlage 66
4052 Basel
Switzerland
Tel. +41 61 683 77 34
Fax +41 61 302 89 18
www.mdpi.com

Energies Editorial Office
E-mail: energies@mdpi.com
www.mdpi.com/journal/energies